USING EXCELERATOR® FOR SYSTEMS ANALYSIS & DESIGN

JEFFREY L. WHITTEN, MS, CDP
LONNIE D. BENTLEY, MS, CDP
Both at Purdue University

1987

IRWIN

Homewood, II 60430
Boston, MA 02116

Dedications

To George McNelly, a consistent source of inspiration to our scholarly and creative endeavors. Thank you for creating the environment in which the Computer Technology Department can excel.

To Chris Grejtak, who made it possible for Purdue to become a leader in Computer-Assisted Software Engineering education. Thank you for your foresight and long-time support.

- Jeff and Lonnie -

Editor	Susan A. Solomon
Editorial Assistant	Lisa Donohoe
Copy Editor	Sally Schuman
Cover Designer	Nancy Benedict

FIRST EDITION
Copyright©1987 by Times Mirror/Mosby College Publishing
A division of the C.V. Mosby Company
11830 Westline Industrial Drive, St. Louis, MO 63146

Printed in the United States of America

Library of Congress Cataloging-in-Publication Data

Whitten, Jeffrey L.
  Using Excelerator For Systems Analysis and Design.

  Includes index.
  1. Excelerator (Computer program) 2. Systems engineer-
ing--Data processing. I. Bentley, Lonnie D. II. Title.
TA168.W455 1987   620.7 2 02855369   87-1957
ISBN 0-8016-5462-9

90 PC 6 5 4 3 2 1

CONTENTS IN BRIEF

TABLE OF CONTENTS

PART THREE: PUTTING THE POTENTIAL TO WORK FOR YOU   121

Lesson Seven: Maintaining the Project Data Dictionary   122

Lesson Eight: Analyzing Documentation Quality Using EXCELERATOR   150

Lesson Nine: Packaging Documentation Using EXCELERATOR   172

Lesson Ten: Where Do You Go From Here?   193

FOREWORD

Over the past decade, systems development professionals have been bombarded with a wide range of new tools, techniques, and methodologies. Each one has claimed to solve the applications development backlog, a problem that has menaced us since the advent of the first computer system. What has become increasingly clear to many of us is that there is no single, easy solution. There is no resolution to the applications backlog or to the more serious problem of our failure to provide systems that meet the real needs of the end-user.

There is one approach that can make a major difference: to understand the user's requirements better at the beginning, then design the system in conjunction with the user and allow the user to simulate the system. Though this may sound like applications prototyping or fourth generation languages, it can best be accomplished by using the new "power tools" designed to support the techniques of structured systems analysis and design. EXCELERATOR is one of these tools.

As Jeff Whitten and Lonnie Bentley will discuss in this book, the new combination of tools and techniques finally allows us to deal with the issues of designing and delivering systems. Possibly now we can deliver systems that are on time, that are closer to budget, and, most importantly, that meet the users' needs. But to ensure the successful introduction and use of these tools and techniques, it is critical that we educate today's -- and tomorrow's -- systems development professionals.

Index Technology feels that colleges and universities should have access to state-of-the-art software products, so that their students will function effectively in the marketplace upon graduation. To that end, we have worked with a number of schools over the last three years, to help them incorporate EXCELERATOR into their systems analysis and design courses. One of the most successful of these relationships has been with Professors Jeff Whitten and Lonnie Bentley of Purdue University. Their use of EXCELERATOR in the classroom and in student projects has helped Index Technology enhance the product to meet the practical needs of the systems analyst.

We are extremely pleased to have had the opportunity to work with Jeff and Lonnie and with Susan Solomon of Times Mirror/Mosby on this text. We feel that the book will help systems analysts learn how to use CASE tools to build better systems, more quickly and more effectively.

Our thanks to Jeff, Lonnie, and Susan.

Chris M. Grejtak
Vice President
Index Technology Corporation
February 1987

PREFACE TO STUDENTS AND INSTRUCTORS

The Intended Audience for this Book

Using EXCELERATOR for Systems Analysis and Design is intended to support a course on EXCELERATOR *or* to complement a systems analysis and design or software engineering course in computer information systems development. The course may be taught at either the undergraduate or graduate level in colleges and universities, and for continuing education in industry. We recommend that the student be familiar with or studying the tools and/or techniques of the *structured systems analysis and design* methodology. At the very least, the student should be studying data flow diagrams, data dictionaries, and data(base) modeling. Familiarity with the concept of prototyping may also prove useful. The book can be used within computer science, computer information systems, business data processing, and software engineering curricula or environments.

Why We Wrote This Book

We have always been fascinated with the possibilities of using the computer to accelerate the productivity of computer-based systems development. We have also long been frustrated by the quality assurance problems associated with modern systems development methodologies. Indeed, methodologies such as *structured systems analysis* offered hope in the form of rigor, but frustration in the form of large specification documents that were difficult to prepare and impossible to maintain. This book introduces renewed hope!

Because demand for computer applications has increased faster than the supply of computer professionals, industry is now turning to computer-assisted tools to get more quantity and quality out of data processing professionals. This trend will likely accelerate since the supply and demand problems are growing more desperate with time. Industry is looking for graduates who are trained in computer-assisted development techniques. Industry is also looking to retrain their existing staffs to use the new technology. These reasons motivated us to write this book.

EXCELERATOR, offered by Index Technology Corporation (InTech), represents the best-selling product for computer-assisted systems development. This fact, coupled with InTech's willingness to work with post-secondary educational institutions, motivated us to build this tutorial around their product.

We have great interest in the EXCELERATOR concept and product. However, we have also noted, with great concern, that many EXCELERATOR users fail to fully exploit the powerful capabilities of the package. For instance, many users view EXCELERATOR as a much-needed graphics tool, but never use the powerful relationships among graphs, graph objects, and detailed dictionary specifications. Many users fail to exploit the power of explosion and the analytical aspects of the product. Our objective is to reduce the learning curve associated with exploiting the powerful capabilities of EXCELERATOR as it can be applied to popular, modern methodologies.

Why We Think You Should Consider This Book

We believe you will find this book unique and should consider it for your course because:

o   *This tutorial is up to date.* It is based on Version 1.7 but it can also be used with version 1.8. Furthermore, because we use EXCELERATOR extensively within our own courses, we are highly motivated to keep the material up to date when InTech releases new versions.

o   *A demonstration case places the exercises into the context of a typical systems development project.*

o   *The step-by-step approach easily guides the student through the learning process.* We also explain why the steps and our recommendations are important. Our recommendations and preferences are based on our extensive experience with the product. We have used EXCELERATOR since Version 1.0.

o   *A detailed art program accompanies the tutorial.* This art program includes appropriate screen-by-screen sequences and sample outputs.

o   *Index Technology, as well as professors who use* EXCELERATOR, *was involved in the review of this tutorial.* We have enjoyed a long-time and good relationship with the InTech. The Foreword to the tutorial was written by InTech's Vice President of Sales and Marketing.

o   *Index Technology is making a free* EXCELERATOR *software offer to educational institutions.* This is the full-function product, not a limited version. See the next section for details.

Free Software Offer from InTech

Any educational institution that requires students to use the Whitten/Bentley/Ho textbook, *Systems Analysis and Design Methods*, 1986, can receive a free copy of EXCELERATOR from Times Mirror/Mosby if the institution meets the following conditions:

1.   Students must *buy* Whitten/Bentley/Ho, *Systems Analysis and Design Methods*.
2.   The college or university bookstore purchases this required textbook from Times Mirror/Mosby.
3.   Ten or more textbooks are ordered from Times Mirror/Mosby.
4.   The educational institution has the computer equipment necessary to operate EXCELERATOR.
5.   The educational institution agrees to follow the license agreement and other stipulations included in the Index Technology Educational Grant Program.
6.   An authorized person from the educational institution completes and submits an official free copy application form to Index Technology Corporation.

This is not a limited version of the product. Contact your Times Mirror/Mosby representative for an application form or call (800)-325-4177 and ask for your sales service representative.

Acknowledgements

We are indebted to many individuals who have contributed to the development of this book. Most of all, we want to thank Index Technology Corporation for their cooperation, support, patience, and endless enthusiasm. Within InTech we offer special thanks to Karen Roman, Judith Vanderkay, Robin Elliot, and Susan Martin, all of whom contributed to the review of the manuscript. Additionally, we thank Chris Grejtak, Vice President of Sales and Marketing, not only for writing the Foreword, but also for his support when we sought to make EXCELERATOR and CASE technology an integral part of our curriculum.

We would also like to thank our academic reviewers: LaVon Green, Purdue Calumet; James Westfall, University of Evansville; and Allen P. Gray, Loyola Marymount University.

Once again, we are indebted to the editorial staff of Times Mirror College Publishing. Your patience and efforts are appreciated, and we continue to look forward to our future joint endeavors.

We would be remiss if we didn't offer a class action apology to all those writers who have ever written tutorials or manuals for software products. Like many others, we have criticized them for a difficult and thankless job. We thought this type of book would be easy. It wasn't. But is was fun, and the end result is equally satisfying. These manuals are harder than textbooks! We tip our hats to those who write software manuals for a living!

Of course, we assume full responsibility for any inadequacies or errors in this book. We plan to keep it up to date to reflect new versions and improved pedagogy; therefore, please address comments, criticisms, and suggestions to us in care of Times Mirror/Mosby, 4633 Old Ironsides Drive, Suite 410, Santa Clara, CA 95054.

Jeff Whitten
Lonnie Bentley

GUIDELINES FOR IMPLEMENTING THE TUTORIAL ENVIRONMENT

General Prerequisites

These comments are addressed to the EXCELERATOR course instructor. To help you set up an ideal EXCELERATOR environment for this tutorial, we make the following assumptions:

o   You have correctly installed EXCELERATOR on your workstation(s) and have successfully logged on to the product (probably under the default user ID, *user*, and the default password, none (the Enter key).

o   The mouse hardware and software have been correctly installed.

o   You are running Version 1.7 of EXCELERATOR. If you are running Version 1.6 or earlier versions, your students may have trouble with keyboard assignments, especially the use of the Enter key, tab key, F3 function key, and F4 function key. InTech greatly improved the functionality of keyboard assignments in Version 1.7.

o   You have not altered the product with CUSTOMIZER, an add-on product sold by InTech.

o   You have some basic familiarity with how to use the mouse to SELECT and CANCEL commands. If not, Skim through Lesson 2 before proceeding.

o   You have a general working knowledge of PC/DOS or MS/DOS.

If you meet all of the above conditions, you are ready to set up the tutorial environment.[1] In this preface, we will make recommendations for setting up your environment. Keep in mind, you may want to alter our recommendations to suit your individual needs.

System and Project Managers

After installation, EXCELERATOR defaults to one initial user account (called *user*). There is no default password for that account. The account privilege is *system manager*. By default, a system manager also has *project manager* access privilege. These two important terms are defined as follows:

o   **System Manager.** A system manager account can gain access to the **System Manager** facility within **HOUSEKEEPING**, a Main Menu option. Through this facility, the account can inspect, add, delete, or change defaults for *all* other user accounts. In most cases, you should have only one system manager account for all workstations. This is especially true in the academic environment. Consequently, we recommend that you add a password for the *user* account ID as follows:

1.   SELECT HOUSEKEEPING.
2.   SELECT System Manager.
3.   SELECT Users
4.   SELECT Modify.
5.   Type *user* and press the Enter key.

---

[1]   Experienced EXCELERATOR environment managers may have already set up class or student accounts. If so, read this section to compare your environment with our recommendations and make changes as desired.

6.   SELECT **Password.**
7.   Type your new password and write it down someplace.

This procedure ensures that only you can access the *user* account to create and delete EXCELERATOR user accounts.

o   **Project Manager.**   A *project manager*, using InTech terminology, is an account that is authorized to access the **Project Manager** facility within **HOUSEKEEPING**, a Main Menu option.   A project manager can create, delete, and change defaults for EXCELERATOR projects. This includes establishing which user IDs can access which project accounts.

In the EXCELERATOR environment, every unique project name is given a unique DOS subdirectory under \EXCEL. In the academic or training setting, this can become unwieldy because too many project accounts can quickly exhaust available disk storage capacity. We recommend that for this environment, the system manager also be the only authorized project manager. Student accounts should not usually be given the *project manager* privilege.

All accounts that do not have *system manager* or *project manager* privileges are said to have a *user* privilege (the default for all new accounts). We recommend this default for all student accounts. We will discuss how to set up accounts and projects below.

Setting Up the Printer Configuration

If your system is connected to a printer or printers that are supported by InTech, you need to configure for those printers. Use the following simple procedure:

1.   SELECT **HOUSEKEEPING.**
2.   SELECT **System Manager.**
3.   SELECT **Configuration.**
4.   SELECT **HARDCOPY DEVICE 1.**
5.   SELECT your printer from the supported list. If you have no printer, we suggest that you SELECT QMS so that your display can be as compatible as possible to the environment we used to prepare the tutorial.
6.   SELECT **Narrow** or **Wide** carriage.
7.   SELECT the proper port.
8.   If prompted, SELECT an appropriate speed for transmission to your printer (serial printers only).
9.   Press the F3 function key to save your configuration.

Setting Up User IDs (Accounts) and Projects

Setting up accounts and projects is not difficult; however, some preliminary decisions need to be made. First, how many students (or student teams) will be using the workstation(s)? For the typical academic setting (many students per station(s), you can go one of two routes:

o   Set up separate accounts and projects for each student (or student team).

o   Set up a single account and project to be shared by all students (and teams).

If you choose the former, students should be informed to carefully monitor disk space usage. If the disk becomes almost full, they must know how to delete (via DOS) a project subdirectory(ies) to create space. And because students may be forced to perform such deletions, the importance of backing up their own projects becomes imperative!

We chose to go the single account direction for the academic environment; however, this too creates some overhead, this time for the instructor. We find it simpler to manage for the typical academic setting. If you create a shared account structure, the following six files should be copied into a DOS directory that you create (e.g. \excel\backup):

All files with the .IDX extension
All files with the .DAT extension

These files are crucial to the startup of EXCELERATOR (without them, you get the error message 50.2). We use the following solution. Make a copy of the six files in a directory called \excel\backup. Then, create a startup batch file (we call it GOEXCEL.BAT) with the following commands:

```
cd \excel
copy \excel\backup\*.* \excel_____
excel
del \excel_____\*.*
```

The blank contains the name of the project subdirectory to be shared by the students (we call it *project*). The first statement changes to the EXCELERATOR executables directory. The second statement copies the essential data and index files into the shared student account. The third statement starts EXCELERATOR. After the student exits EXCELERATOR, the fourth statement deletes the project directory to ready the shared account for the next user (obviously, it is imperative that the student use EXCELERATOR's Backup facility, covered in Lesson 3, to save the work to diskette). Students should be told to always start this environment with the command GOEXCEL instead of EXCEL.[2]

Once you've made the above decisions, you set up accounts using the following procedure (starting from EXCELERATOR's Main Menu):

1.   SELECT HOUSEKEEPING.
2.   SELECT System Manager.
3.   SELECT Users.
4.   SELECT Add.
5.   Type an Account Name (also called an account ID) and press the Enter key.
6.   This takes you to a User Record screen. Type in a password and press the Enter key.
7.   If you want a privilege other than *user* (reminder: not recommended for academic settings), type the appropriate letter code.
8.   Press the F3 function key to save the account.
9.   If you want multiple user accounts, repeat steps 4-8.

Now that you have established user accounts, you should set up projects and link them to accounts. From the Main Menu, use this procedure:

1.   SELECT HOUSEKEEPING.
2.   SELECT Project Manager.
3.   SELECT Add.
4.   Type a project name (also called a project ID) and press the Enter key.
5.   This takes you to a Project Manager screen. Do the following:
      a.   Type a SHORT NAME. This is the name that will be given to the project subdirectory (if you are using the batch file startup technique described earlier, this short name must match the name you recorded in the blank.
      b.   SELECT USERS. Type in a user account name (ID) that will use this project directory. If you use our shared directory idea, there will only be one user -- that's why this academic setup is easier to manage. Press the Enter key. Type in the ACCESS code. It should be *M* in order for the students to complete our tutorial.

---

2   Alternatively, experienced DOS users could rename the EXCELERATOR startup command from EXCEL.EXE to EXCELST.EXE, rename the batch file to EXCEL.BAT, and change the third statement to excelst. This would permit the student to start EXCELERATOR with the standard command, EXCEL (executing the batch file).

c. **SELECT DATA FLOW DIAGRAMS.** Important! EXCELERATOR allows you to establish the data flow diagraming symbol set. The default, used for figures throughout this book, is *G* for *Gane & Sarson*. This default uses rounded boxes for the process symbol. If you prefer the *Yourdon* symbol set, which uses circles for the process symbol, type *Y*.

d. **SELECT DATA MODELING.** Important! EXCELERATOR allows you to establish the data modeling symbol set for entity-relationship diagrams. The default, *C*, uses the Peter Chen symbols. If you refer the *Merise* symbols, Type *M*. The figures in this book are based on the default setting.

e. Press the F3 function key to proceed to the next screen.

f. This screen establishes default object and font sizes for all graphs. Important: These values affect the maximum display size for a label on a graph. We recommend that you consult the EXCELERATOR Reference Guide and SELECT the smallest object and font size possible for your supported printer. For the QMS laser printer, used to print the graphs you will see in this tutorial, the settings were:

FONT SIZE ---> 2 SMALL
OBJECT SIZE ---> 5 MEDIUM

The font and object sizes also affect how many objects can be placed on one page. Choose your sizes carefully.

Warn your students that their graphs may not print entire labels as shown in our graphs, especially since we used one of the highest resolution printers that InTech supports.

g. **SELECT EXIT** to save your project defaults.

6. If you are setting up multiple projects, repeat steps 3-5.

Linking Your Word Processor and Project Manager Into EXCELERATOR. One of the later lessons in the tutorial demonstrates EXCELERATOR's direct link to your word processor and project management software. If you want to take advantage of this lesson, you must properly link your licensed copy of said packages to EXCELERATOR. Assuming you have already installed either or both of these packages on your hard disk, you need the EXCELERATOR Utilities diskette to execute the link. Place the diskette in Drive A. Type **PATHMOD** and press the Enter key. Follow the directions on the screen to tell EXCELERATOR where to find your word processor's (and project manager's) executable files and to identify the startup file. Then **Exit**. Those packages are now callable from EXCELERATOR's **DOCUMENTATION** menus.

Before you assign Lesson 9, we recommend that you read that lesson and prepare the indicated word processing files for your students' use.

What About Multiple Workstation Environments? If your training environment uses multiple workstations, we recommend that you set up one workstation as desired (complete with accounts and projects). Then, backup that workstation's root, \excel, and all project subdirectories. Finally, restore the directories to the other workstations. This way, you won't have to separately install and set up each workstation. But remember, you can only run EXCELERATOR on workstations equipped with a licensed security device. Also, be careful not to create illegal backups of other software packages.

Your environment should now be ready to accommodate your students use of this tutorial.

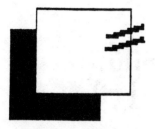

Part One:
Excelerator Concepts
and Fundamentals

Shoes for the cobbler's children. You've all heard the story of the cobbler whose own children had no shoes. That situation is not unlike that faced by systems analysts -- all that wonderful technology and here we have systems analysts whose principle tools are paper, pencil, and flowchart template . . . until now.

EXCELERATOR and similar tools have been likened to new shoes for the cobbler's children. And what shoes they are! EXCELERATOR is a workbench - a collection of automated systems analysis and design software tools. In Part One, EXCELERATOR, the best-selling systems analysis productivity software package, is introduced.

Part One consists of two lessons. The first lesson doesn't require access to EXCELERATOR. It introduces you to the past, present, and future of a concept that will very likely change the way you do systems development over the next decade and longer -- Computer-Aided Systems Engineering (CASE). Don't skip this lesson. It will help you understand how powerful and remarkable EXCELERATOR really is. Many professionals who have access to EXCELERATOR have little concept of the full potential of the tool. The purpose of this lesson is to help you exploit that potential and significantly enhance your career.

In the second lesson you begin your hands-on training with EXCELERATOR. You will learn how to log on, how to set some important system parameters, how to select functions and commands, how to save your work, and how to exit EXCELERATOR.

Put away your flowchart template. Toss your eraser. You are about to enter the new generation of systems analysis and design!

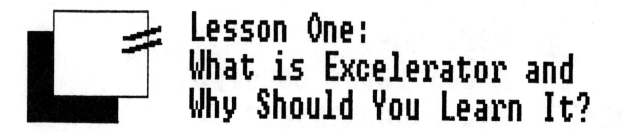

Lesson One: What is Excelerator and Why Should You Learn It?

The Challenge of Today's Data Processing System's Professional

Productivity! If ever there was a watchword for today's business mood, that word would be *productivity*. Why? We believe that management`s concern with productivity is a result of our most recent economic recession. The last recession was different from earlier ones. How so? For the first time in history, we saw greater layoffs in the white-collar workforce than in the blue-collar workforce. True, the blue-collar layoffs got more attention in the news media; however, blue-collar workers were not the most affected.

Why so many layoffs in the white-collar workforce? You could legitimately argue that white-collar workers make up the majority of the workforce (since 1957 -- surprised?). We should have expected that the white-collar workforce would experience the greater number of layoffs, right? Consequently, we can dismiss the layoff numbers as demographically valid, correct? Not necessarily.

In the aftermath of the recession, we have noticed another trend. There has been a permanent decrease in the number of individuals holding middle-level management positions. This decrease is not due to an increase in the number of individuals retiring from these positions. Nor is it due to a decrease in the number of individuals seeking such positions. So what is causing this trend? A new emphasis on productivity! Corporate America decided that management was too fat. The infrastructure we call middle management was largely created to generate, manipulate, and disseminate information between the operational levels of organizations and executive management. These are tasks that the computer can do much more efficiently than the middle manager.

What has all this got to do with you, the systems analyst (or programmer, programmer/analyst, systems engineer, management consultant, data administrator, or the like)? Well, the problem is that the middle managers didn't see their fate coming. They assumed that they were indispensable. This apathy could have been predicted by examining history. Consider the agricultural age. Farmers assumed that the demand for food would provide sufficient job security. But most farmers were eventually replaced by more productive farmers. Then came the industrial age. Throughout that era, we have seen increasing percentages of blue-collar workers displaced by manufacturing mechanization (a continuing trend in light of robotics and computer-assisted manufacturing), again with an emphasis on increased productivity. Many of those workers also thought themselves indispensable. And in this information and service-oriented economy, we are seeing middle-management ranks reduced by more productive computer-based information systems. Who is next? Possibly you! That's right; you might be next.

Crazy? If you value your career, keep reading. We know you've heard the predictions. The numbers vary but at least one noted consultant and lecturer has estimated that if the demand for new applications continues to grow at the current pace, we will need 28 million programmers by the early 1990s. How many programmers do we have today? A little over a quarter of a million. You don't need a calculator to figure out that we can't meet that kind of programmer demand. But before we get too comfortable with the implied job security in this situation, read on.

How do we meet the current and projected demand for applications? Perhaps we concede that the application demand cannot be met. That would be a grievous error. We just saw that economic history suggests that failure to meet demand is frequently perceived to be a productivity problem and that the economy will find a way to replace the less productive elements in order to meet demand. We have become those elements. We must become more productive. Those who do improve their productivity will survive. Those who don't may (or will) suffer the same fate of the early farmers, industrial blue-collar workers, and middle managers.

Still not convinced? It has already started to happen. The first productivity target was (or is) programmers. We are seeing increased usage of fourth generation languages and applications generators. At a minimum, these languages have improved productivity by allowing us to prototype systems instead of spending excessive effort on paper specifications that proved incomplete, inconsistent, and inaccurate. As these languages get more efficient, the prototype is less frequently discarded. Instead, the prototypes are completed and placed directly into operation (bypassing the traditional approach of reprogramming the system in third generation languages such as COBOL). End users are now using these languages to develop their own systems (we'll ignore discussion of the control and integrity issues of end-user programming). We are also seeing renewed interest in automatic code generators. If you are not familiar with these programming productivity trends, we suggest you read James Martin's classic book, *Application Development Without Programmers* (Prentice-Hall, 1982). It's a real eye-opener.

The next target for productivity enhancement is the systems analyst, database specialist, expert systems specialist, et al. Unfortunately, the goal of improving productivity in systems work has been a two-edged sword. On the one hand we are asked to be more *productive*. On the other hand, we are frequently criticized for not producing *quality* systems. Historically, we have viewed productivity and quality as equal and opposite forces. In other words, efforts to improve quality (e.g., *Structured Analysis*) have resulted in reduced productivity (and vice versa). As conceptualized in Figure 1-1, we find that efforts to increase productivity tend to decrease quality and vice versa. Indeed, data processing and systems development managers find themselves walking a tightrope, trying to find the optimum balance between productivity and quality. But it doesn't have to work that way. We can eat our cake and have it too -- enter Computer Aided Systems Engineering (CASE). Improvements in productivity and quality are the prime goals of CASE.

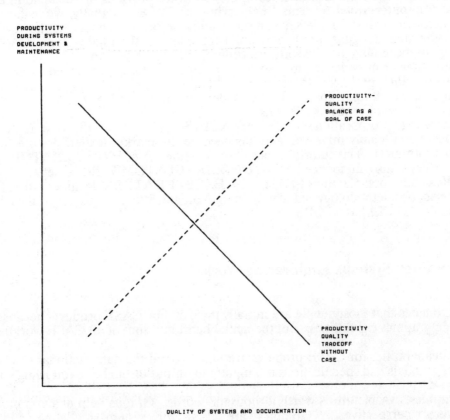

Figure 1-1

Data Processing Productivity/Quality Tradeoff

Computer-Aided Systems Engineering (CASE): CAD/CAM for the Systems Specialist

The concept of using computers to assist with systems analysis and design is not new. It dates back to the early-to-mid 1970s. The most successful product was *Problem Statement Language/Problem Statement Analyzer (PSL/PSA)* as developed under Dr. Daniel Teichrowe at the University of Michigan. This product ran on large mainframe computers. PSL was a language for specifying user requirements. PSA was an analyzer that evaluated the quality of the requirements statement. The ultimate goal of the project was to eventually generate code from the requirements statement. At the time, PSL/PSA was capable of bringing computers to their knees. Few companies could afford to release enough PSL/PSA access time from their production machines (the computers running their applications programs). Neither could they afford dedicated PSL/PSA computers. The concept of computer tools for systems analysts was still, essentially, an idea ahead of its time.

Then came the microcomputer. The microcomputer offered the option of distributing support for systems analysis and design activities from the production computers to dedicated systems analysts' microcomputer workbenches. You are probably familiar with common microcomputer software tools such as word processors, spreadsheets, and data managers. Such tools offer some productivity improvement potential for systems analysts. We can write reports more efficiently with word processors. We can use spreadsheets to improve estimating and cost/benefit analysis. Today's microcomputer database management systems have evolved into full-function fourth-generation languages and applications generators that can replace paper design specifications with functional prototypes. But these capabilities only scratch the surface.

Today's analyst needs system modeling tools that support systems analysis and design graphics and help the analyst manage and control the enormous volume of details that are associated with all but the smallest projects. A new generation of microcomputer software provides these capabilities and more. We call this software **Computer-Aided Systems Engineering**, or **CASE**. Actually, we've taken some editorial license with the acronym. Most people say computer aided *software* engineering; however, we are *systems* analysts, not *software* analysts. Software is only one component of the total system. CASE assists with the analysis and design of more than just the software components of the system.

CASE automates or provides automated support for many tasks performed by analysts. This is very similar, in concept and practice, to *Computer Assisted Design/Computer Assisted Manufacturing (CAD/CAM)*. In fact, the ultimate goal for CASE is identical to that of CAD/CAM: to automate totally the design and implementation of products (systems).

Index Technology Corporation's (InTech) EXCELERATOR was the first IBM PC-based CASE product. It has been continually improved and has become the market leader, despite the introduction of several competitive products. This tutorial is based on Version 1.7 of EXCELERATOR. With the release of Version 1.7, InTech also announced *CUSTOMIZER*. CUSTOMIZER, a separate product, permits EXCELERATOR's client organizations to customize EXCELERATOR's documentation facilities to match data processing and/or methodology standards in the organization. This tutorial is based on a non-customized version of EXCELERATOR.

How Computer-Aided Systems Engineering Works

Our experience indicates that most people are initially intrigued by CASE products because of their systems analysis and design graphics capabilities. But the actual heart and soul of all CASE workbenches is the data dictionary.

Functional capabilities for CASE products are built around the data dictionary. This is illustrated in Figure 1-2. System models and specifications are input and output through the following common facilities:

o   **Graphics.** A picture is worth a thousand words. Graphs help analysts model systems from several perspectives. Graphics are used for system modeling and decomposition. EXCELERATOR provides for the following graph types:

* Data Flow Diagrams (Yourdon and Gane/Sarson symbol sets)
* Structure Charts (Yourdon/Constantine)
* Structure Diagrams (Michael Jackson)

4

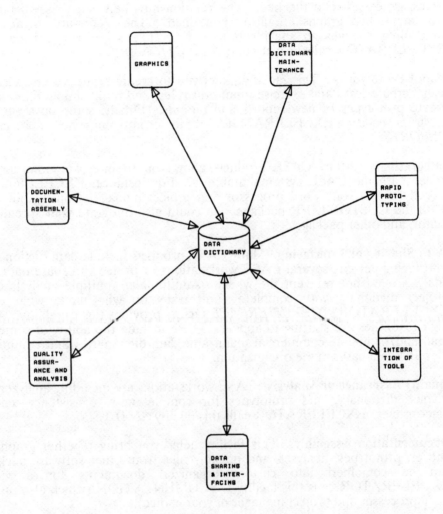

Figure 1-2

CASE Facilities and Architecture

5

    *   Entity-Relationship Data Models (both Chen and Merise symbol sets)
    *   Data Model Diagrams (based on Bachman techniques)
    *   Presentation Graphics (a superset of ANSI system and program flowcharts)

Graphs can be exploded or leveled into more detailed graphs. The relationships among these graphs are maintained in the data dictionary.

o   **Data Dictionary Maintenance**. The data dictionary is used to capture, modify, and control details about data flows (both input and output), data models, records, data elements, table of codes, processes, and the like. The relationships between these dictionary items and the aforementioned graphs are also maintained. The dictionary provides a mechanism for controlling changes to specifications. It is like your own private data administrator. EXCELERATOR calls its dictionary *XLDICTIONARY*.

o   **Rapid Prototyping**. This facility allows you to create input screens, test those screens with users, create test data, exercise input edit rules, and build simple files. It also allows you to create prototypes of new reports and outputs. Finally, some prototypers generate limited program code. EXCELERATOR's rapid prototyping facility is called *SCREENS & REPORTS*.

o   **Integrating**. Many CASE products allow you to integrate other productivity packages underneath the CASE system's umbrella. For instance, EXCELERATOR allows you to directly access your word processor and project manager from within EXCELERATOR. With the CUSTOMIZER package, you could also integrate your spreadsheet, terminal emulator, and other packages.

o   **Data Sharing or Interfacing**. Many CASE products include data dictionary sharing facilities. Interfacing permits several CASE workstations or projects to share data dictionaries. Thus, analysts need not reinvent the wheel. Additionally, multiple analysts can share the same project dictionary with complete audit trails on who made what changes and when. EXCELERATOR's *XLD INTERFACE* facility provides this capability. Advanced data sharing capabilities are starting to appear. These include the ability of some CASE products to read and/or update commercial mainframe data dictionaries that administer the production files and data bases in the organization.

o   **Quality Assurance or Analysis**. CASE workstations are including analyzers that evaluate your graphs, dictionary, and prototypes for completeness, consistency, common errors, and inaccuracies. EXCELERATOR calls this facility *ANALYSIS*.

o   **Documentation Assembly**. This facility helps you bring together graphics, data dictionary entries, prototypes, analyses, and possibly files from other software packages. These items can be combined into custom organized documents for a variety of sources. EXCELERATOR calls this facility DOCUMENTATION (which also provides access to the word processor and project manager of your choice).

The end result of CASE is a collection of integrated productivity tools that can, when properly and fully applied, significantly enhance virtually every data processing professional`s productivity.

What Potential Benefits Can Be Realized From CASE?

Like any other tool, CASE can be misapplied or not used to its fullest potential. Indeed, the purpose of this tutorial is to help you fully realize the following CASE benefits:

o   Increased systems analyst productivity. Estimates of improved productivity range from 35 to more than 100 percent. These estimates are based on exploiting much of CASE's potential -- more than just graphics.

o Higher-quality systems documentation. There has been an increased willingness on the part of users to maintain documentation.

o Higher-quality systems development. The ability of CASE to perform quality assurance pays dividends.

o Systems development methodologies that work. CASE makes it possible to realize the benefits of a specific systems development methodology or integrate the capabilities of multiple systems development methodologies.

o Easier to create and enforce standards. Standards increase the probability of less expensive systems maintenance.

o Increased analyst/user community morale. Analysts will likely feel better about their jobs. It is even conceivable that analysts will gravitate toward organizations that provide such state-of-the-practice tools for their analysts.

How Do You Sell CASE to Management?

You may be wondering how to sell CASE to management. This is especially likely if you are a student looking to take the skills you learn in this book into the working world. The concept is not easy to sell. Despite better productivity -- despite all the aforementioned benefits -- when the chips are down, all good managers want to talk money, specifically cost/benefit analysis. CASE isn't cheap. The first reaction to the price of CASE is frequently, "My shop can't afford to do that!" We would like to counter that your shop cannot afford *not* to do it. And we can prove it with a very simple financial analysis.

This analysis is based on worst-case assumptions that are designed to make the CASE concept *infeasible*. That's right. We said infeasible. Our approach says let's overestimate the costs and underestimate the benefits. If CASE works under these conditions, your firm should experience much greater return on the investment. The analysis works as illustrated in Figure 1-3.

As you see, even with overestimated costs and underestimated benefits, CASE yields a positive return on investment. When you make the numbers reflect quantity discounts, more realistic salaries and overhead, and better, more typical productivity gains, the net present value grows significantly -- even when you figure in maintenance costs. Also, we haven't even figured in the intangible benefits of higher-quality systems, user satisfaction, analyst morale, and improved documentation. Consequently, we reassert our claim, "you can't afford NOT to do it!"

One thing's certain. You shouldn't march into management with a proposal to buy too many workstations at once. One or few workstations is a better option. It is equally important to make the product prove its value. That is the only way to multiply the number of workstations. That brings us to our next question ...

How Do You Make Case Work in Your Shop?

In other words, how do you become a CASE activist and make CASE a career booster? These questions are equally pertinent to college students and to practicing systems analysts. We offer the following suggestions:

o Students frequently ask us how they can find organizations who use EXCELERATOR or other CASE products. They would like to seek employment with such organizations. One approach is to contact user groups; the CASE vendor can get you in touch with their customers. Another way is to prominently feature the CASE product's name or use the terms "Computer-Aided Systems Engineering" or "CAD/CAM for the Systems Analyst" on your resume.

Costs of CASE

The microcomputer needed to run CASE. We'll ($ 6,700)
go with an IBM PC/AT instead of the cheaper but
workable IBM PC/XT (or many compatibles)

The CASE software (9,000)

Total Costs ($15,700)

Benefits of CASE (and How They Were Derived)

Annual cost of a systems analyst $25,000
(we'll use a conservatively low
average salary)

Overhead including benefits,   10,000
continuing education, mistakes,
and the like -- again, the
numbers are probably underestimated

Total employee annual cost   35,000

Estimated productivity en-   15%
hancement (we'll use a con-
servative estimate -- some
actual CASE users estimate
productivity improvements of
35-100%)

Annual productivity value ($35,000 X 15%) $ 5,250

Lifetime Benefits (annual X 5 years) $26,250

Cost/Benefit Analysis

Like good financial managers we'll time adjust our
numbers, assuming the five-year lifetime and 10%
cost of capital.

Time-adjusted lifetime costs   ($15,700)

Time-adjusted lifetime benefits (discounted)   19,900

Net Present Value   **$ 4,200**
NET FUTURE VALUE   **$ 6,766**

Figure 1-3

Cost/Benefit Analysis for CASE

But we would also like to propose an alternative approach. Why not search for companies who don't currently have CASE but who are openly interested in your background with CASE? People make their own breaks in the business world. Make yours! Assert yourself and become the CASE activist or champion in an organization.

o   Choose appropriate pilot projects. Don't select the monster, two-year, multi-analyst project as your first CASE experience. Try a small project of less than one calendar year estimated duration.

o   Consider initially using CASE to document an existing system without the pressures of a systems development project. This gives you a vehicle for learning the product and trying many features.

o   When you first use CASE on a project, commit yourself to using more than the graphics and basic data dictionary. Try to become the first analyst to fully exploit the CASE capabilities. Become the local CASE expert.

o   Join the user group for the CASE product you use. Get management to support your attendance at user conferences so you can learn and share new ideas. User groups will also be a key influence to future CASE improvements.

o   Using CASE techniques will be like learning to drive a car with a manual transmission. While you're learning, you are constantly thinking about the clutch release, RPMs, shift pattern, not stalling the engine, etc. It seems like you'll never learn or get used to it. But eventually you stop thinking about it - you just do it. Early on, it won't seem that you are achieving significant productivity improvements with CASE. That's because you're learning the tool. With each successive project, you'll see your productivity improve. Hang in there!

The Future of CASE ... Don't Get Left Behind!

The future is bright for CASE. Initially, the goal would be to have a CASE workstation on every analyst's and programmer's desk. Dictionaries would be shared on file servers, available to all project team members. The next logical step is to make CASE mobile. We are now starting to see the first generation of IBM PC/AT compatible lap-top portable computers, complete with hard disk. Before long, these lap-tops will be able to support all CASE functions. This will give the analyst a degree of flexibility that is yet unrealized - the ability to take the tool into the user's work area, verifying and maintaining documentation on the fly. Also, it would be conceivable to get the users into a room where walkthroughs could be conducted directly using CASE - no longer constrained by a subset of documentation that has be printed to paper.

And CASE products are going to get much better. They will eventually achieve total automated support of systems development and maintenance. Figure 1-4 depicts what we feel are the future capabilities for CASE products. The facilities will continue to be centered around the data dic-tionary:

o   **Improved Networking.** This will allow multiple workstations serving multiple analysts who are working on the same project to more easily share project data. This capability is more dependent on the vendors of networking technology than the vendors of CASE technology. Networking is already available in many CASE products, including EXCELERATOR.

o   **Data Modeling and DataBase Design Aids.** CASE workstations will give us tools that help normalize data models or check normalization efforts. And they will provide translators to convert conceptual data models into a variety of popular logical data models as supported by commercial databases such as *IMS, IDMS, SUPRA, DB2*, and maybe even the increasingly powerful microcomputer databases like *DBASE III PLUS* and *System V*. Data models will probably remain subject to modification and fine tuning by database experts.

o   **Improved Quality Assurance.** The workstations will eventually be able to more thoroughly analyze your specifications to find problems and inconsistencies.   Analysis will include

9

recommendations that offer hints on how to improve the specifications. This capability borders on ...

o   **Artificial Intelligence.**   CASE workstations will include rule-based analysis and design that reflect the knowledge and expertise of individuals.   In other words, the workstation will provide not only error *detection* but error *correction*.

o   **Additional Data-sharing Bridges.** Eventually, all good CASE products must provide bridges to and from all major commercial mainframe data dictionaries, including those which are tied to specific database management systems.

o   **Code Generators.**   Some current CASE workstations, including EXCELERATOR, can generate data divisions directly from specifications. We see future workstations generating logic and processing code as well. Initially, the workstations will probably generate skeleton code, leaving completion of the programs to programmers. Code may be in third-generation, procedural languages such as COBOL, ADA, and C. Ultimately, code will be in fourth-generation non-procedural languages such as FOCUS and even DBASE. Programs generated in non-procedural languages would be easier to maintain. It is even conceivable that maintenance could be accomplished directly through CASE specifications for the final system.

It is not clear whether code and/or application generation will be built into the CASE workstations or provided via bridges to mainframe generators. It will probably work both ways.

o   **Maintenance Through the Specification.**   Most DP managers and professionals reluctantly admit that systems documentation is rarely maintained through the system's operational lifetime. The pressures for new and modified systems prevent that luxury. But consider this prospect. A system is generated from the CASE workbench. If that becomes possible, then why not maintain the system via the CASE specification? That way we eliminate the classical maintenance approach *and* maintain up-to-date documentation. And if this concept can be realized, the next item may also be possible ...

o   **Specification Generators.** Consider the idea of decompiling existing software backwards into CASE specifications. The specifications could then be modified and fed into code generators or application generators that create improved systems. All this and no programmers!

How These Lessons Are Organized

All of the lessons in this tutorial are organized as follows:

o   Lessons begin with learning *objectives* under the heading WHAT WILL YOU LEARN IN THIS LESSON?
o   Lessons are divided into *exercises*. Exercises should be completed in sequence. We suggest that you complete exercises in single work sessions until you have learned how to back up your work (Lesson 3).
o   Exercises consist of several *steps*. We recommend that you perform the steps exactly as we suggest. Otherwise, you may find yourself in a position where you can't try something that we've specifically set up for you.

Throughout this tutorial, we will use the term *work session* to describe one continuous EXCELERATOR session, after which you exit EXCELERATOR and leave the workstation. Most of the time, you will be able to quit the work session, save your work, and then pick up where you left off when you start the next work session. Until you know how to do *all* of these things, we suggest you end a work session only at times we suggest. We wouldn't want you to lose your work.

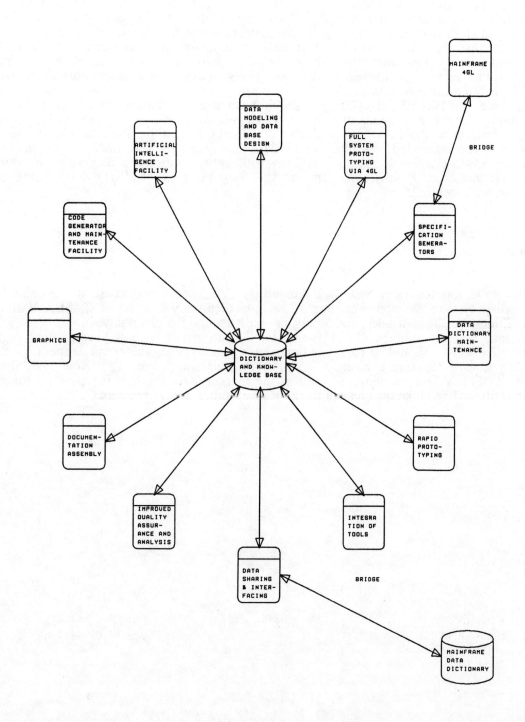

Figure 1-4

Future Architecture of CASE

Prerequisites to the Lessons and Exercises

There is only one technical prerequisite to the tutorial. You must have access to EXCELERATOR software installed on a personal computer. This tutorial does not cover installation. Installation is performed by an individual that InTech calls the *system manager*. The system manager must establish accounts and passwords. The system manager also needs to set up certain system defaults. Finally, the system manager (or *project manager*) must also establish projects for the accounts. Before you start, confirm that the system manager has performed these tasks. The system manager should consult the *Guidelines for Setting Up the Tutorial Environment*.

As a student of EXCELERATOR, you should have already read Lesson 1. Otherwise, you may not fully understand the tools you are about to learn. Also, you should have completed or be in the process of completing a systems analysis and design course or training program. Some familiarity with structured methodologies is recommended. Familiarity with data flow diagrams is assumed. If you plan to cover the data modeling lesson (Lesson 5), some familiarity with entity-relationship diagrams (or equivalent data modeling tool) is assumed. Finally, some familiarity with the PC/DOS or MS/DOS operating system would be helpful (but not essential).

Conclusion

CASE stands for Computer-Aided Systems Engineering. CASE seeks to increase systems quality of systems while decreasing systems development time. The data dictionary is heart of CASE. Various facilities, including graphics, quality assurance, and others, are built around the dictionary.

You may find the concept of CASE and totally automated systems development frightening. It shouldn't be. Those who should be concerned are those who don't see this trend happening. You've taken the first step toward the future already -- you've learned what's coming. Now take the next step. EXCELERATOR has firmly established itself as the market leader in CASE. Use this tutorial to teach yourself the tools. Then make the concept work on case studies or real projects.

Lesson Two: Getting Started With Excelerator

The Demonstration Case Study

Before we launch into the hands-on lessons, you might want to become acquainted with our demonstration case study. It is not absolutely crucial; however, it may make the exercises more enjoyable.

Soundstage Record Club is a subscription service similar to those you've seen advertised in magazines and newspapers. Customers join Soundstage through special advertisements or through existing member referrals. The advertisements usually dangle a carrot such as "Select any ten records for a penny and agree to buy ten additional records within a two year period at regular club prices." The member referrals offer incentives to both the new member and the member who referred them. The incentives may come in the form of bonus selections or larger-than-normal discounts to the referring member. The new member enters under the same type of agreement previously described.

Club members receive monthly promotions that offer a variety of products including a "selection of the month." They also receive a few special offers each year (closeouts, two for the price of one, and the like). Every offering requires that the customer respond by a certain date, after which the "selection of the month" or "special selection" would automatically be shipped and billed. Customers can also buy alternative selections or additional selections (to the special), chosen from a catalog provided in the promotional package.

After a member fulfills the initial agreement, he or she earns bonus coupons for every additional item purchased. Bonus coupons can be applied to future purchases.

What Will You Learn in this Lesson?

This lesson will get you started with EXCELERATOR. The purpose of the lesson is general familiarization. It is also important to note that you learn what we mean by *SELECT* and *CANCEL*, two instructions that will be repeated throughout this book. Specifically, you will know that you are ready to proceed to Lesson 3 when you can:

1. Log on to EXCELERATOR and find your way to the main menu.
2. Briefly describe the main menu facilities.
3. Use the mouse to SELECT and CANCEL choices from the EXCELERATOR menus. Also, describe other uses of the mouse and keyboard equivalents.
4. Set certain **Profile** options within EXCELERATOR.
5. Learn how to get help from within EXCELERATOR.
6. Exit from EXCELERATOR.

Exercise 2.1   Log On To EXCELERATOR

Logging on to EXCELERATOR is simple. Make sure you have your account name and password handy.
Because EXCELERATOR's log-on procedure is case (not the acronym CASE) sensitive, be sure to note
whether the account name and password are upper case, lower case, or a combination. (Note: in the
preface, we suggested upper case.) Confirm this with your instructor or system manager. Then proceed as
follows:

Step 1.   Turn on the computer. After the computer goes through its warm-up routine as specified by
            the system manager (in something called an AUTOEXEC.BAT file), the system will give you
            a command prompt (unless the system manager has provided for a menu driven system - in
            which case you should ignore the rest of this step). The command prompt *usually* looks like
            this:

                                        C>

            It means that the default disk drive is *C*, the fixed disk. The EXCELERATOR software is
            installed on the fixed disk. The arrowhead is requesting that you enter a command. Since we
            don't know if your system manager has installed a PATH statement to tell the operating
            system where to search for software, we'll respond to the prompt by telling the system exactly
            where EXCELERATOR is. Type *CD \EXCEL* and then press the Enter key. This changes
            the default directory to the directory where EXCELERATOR is stored. Your screen should
            now display one of the following prompts:

                                        C>

                                   C:\EXCEL\ >

Step 2.   Access to EXCELERATOR requires either the installation of a block security device
            (installed on the computer's parallel port) or the availability of a key diskette. Ask your
            instructor or system manager if you need a key diskette to log on to EXCELERATOR. If not,
            go immediately to Step 3. If you need the key diskette, place it in Drive A. If you're using an
            IBM PC/AT, type *DIR: A* and press the Enter Key. The key diskette must remain in Drive
            A: during the session (except during backup and restore).

Step 3.   To access EXCELERATOR, type *EXCEL* and press the Enter key. You will soon see the
            EXCELERATOR banner screen. Press the Enter key. You should now be at the
            EXCELERATOR Log On screen (Figure 2-1).

            If you get the message, *Unauthorized Duplicate*, it means that you either need a key diskette or
            a block security device. See your instructor or system manager.

Step 4.   Remember, the log-on procedure is case sensitive. Type your account name in the proper
            case and press the Enter key.[1] Now type the password. Notice that the password does not
            print. Press the Enter key. If you have correctly typed the account name and password, you
            should see a project name or names appear (similar to Figure 2-2, although the pictured
            account has access to several project accounts). If so, go to Step 5.

            If the system doesn't recognize your account and/or password, you will see an appropriate
            message with an option to try again. If you get this message, type *Y* for "yes" and repeat the
            log on procedure. If you have repeated trouble logging on, the cause could be either
            typographical errors, typing in the wrong case, or an incorrect account/password from your
            instructor or system manager. Contact that individual if the problem persists.

---

1   Beginning with Version 1.7, the F3 function key can be frequently used in most cases where we suggest the use of the Enter key. However, to simplify
         the tutorial, we have adopted the convention of using the F3 function key only to save and exit data dictionary screens.

Figure 2-1

EXCELERATOR Log On Screen

Figure 2-2

Successful Log On Screen With Project Names

Step 5. In EXCELERATOR, every unique project is given a DOS directory. This directory will be used to hold all of your EXCELERATOR work files. Selecting the project is EXCELERATOR's way of switching to the correct DOS subdirectory.

Normally, you will select projects with the mouse. For the time being, use the arrow keys to highlight the desired project and then press the Enter key. This takes you to the Main Menu (Figure 2-3).

You are now ready to proceed to Exercise 2.2. Do not quit your work session. Immediately proceed to Exercise 2.2.

Exercise 2.2   Use the Mouse to Find Your Way Around the Main Menu

In this exercise, you will learn about the main facilities of EXCELERATOR and how to select them. We will concentrate on the mouse selection technique since that input device may be new to you and because it is mandatory to parts of EXCELERATOR.

Step 1. Try not to get ahead of us. First, let's briefly review the Main Menu screen. This menu screen is representative of most of EXCELERATOR's menus. The Main Menu provides access to the following major facilities:

o   **GRAPHICS.** This facility provides access to all graphics tools in EXCELERATOR.

o   **XLDICTIONARY.** XLDICTIONARY is EXCELERATOR's name for their data dictionary maintenance facility. This facility provides access to the dictionary for creating, deleting, or modifying system details or generating subsets of the dictionary for quality assurance analysis.

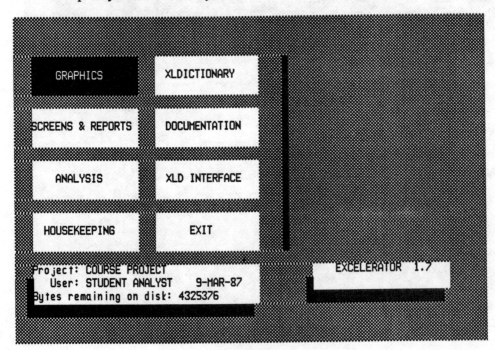

Figure 2-3

EXCELERATOR Main Menu Screen

16

o **SCREENS & REPORTS.** This facility helps you design and prototype screens and reports. After you learn to use it, you may wonder why you would ever do a paper design or COBOL DATA DIVISION again.

o **DOCUMENTATION.** DOCUMENTATION provides access to your computer's word processor and project manager. It also provides a mechanism for combining different graphs, data dictionary entries, screen and report designs, and word processing documents into a single document for printing.

o **ANALYSIS.** This facility provides access to EXCELERATOR's quality assurance routines. We'll learn more about them later in this tutorial.

o **XLD INTERFACE.** XLD INTERFACE provides a mechanism for sharing data between projects and outside DOS directories.

o **HOUSEKEEPING.** HOUSEKEEPING provides access to general purpose functions to set or modify default EXCELERATOR parameters, backup and restore project directories, establish and modify projects (if the system recognizes your account as a project manager account), and establish and modify accounts (if the system recognizes your account as a system manager account).

o **EXIT.** EXIT allows you to exit EXCELERATOR, change to a different project, or change to a different account (assuming you have access to another account).

Menu options are selected in one of three ways:

o the mouse
o the keyboard cursor keys
o alphabetic letters and numbers

We want you to concentrate on the mouse method since several EXCELERATOR facilities require the mouse (e.g., **GRAPHICS**). The mouse is a special input device that may be unfamiliar to you. Many word processors, spreadsheets, and graphics packages allow optional use of the mouse. Our tutorial is based on the Microsoft MOUSE, the most commonly encountered mouse in EXCELERATOR environments. Before we practice, you need to learn some special terminology that we will use to describe mouse functions.

A mouse is a small hand-held pointing device. It fits into the palm of your hand. You move the mouse on a flat surface to move the cursor on the screen. The buttons on the mouse are used to execute and cancel commands. For some people it takes some getting used to. Trust us when we say that it will soon become natural -- you won't even think about it!

IMPORTANT! There are two mouse-related terms that we will use throughout this tutorial (all lessons):

o SELECT. SELECT means that we want you to use the mouse to move the cursor to a menu option or screen location and then press or click the *left* mouse button.

o CANCEL. CANCEL means that we want you to press or click the *right* mouse button. For the CANCEL command there is no need to move the cursor to a location.

Now let's learn to use the mouse to SELECT and CANCEL commands. You should still be at the Main Menu screen.

Step 2. To move the cursor on the screen, move (or roll) the flat-bottom part of the mouse on a flat surface. Try it. Just move the cursor around the screen (without clicking the mouse buttons).

Step 3. SELECT **GRAPHICS**.[2] Remember, we said that SELECT means to move the cursor to the menu option and click the *left* mouse button. The cursor can be located anywhere inside the menu option's box. If you successfully SELECTed **GRAPHICS**, the Graphics Menu (Figure 2-4) should appear on the screen. If the Main Menu still appears or you get an error message, contact your instructor or system manager.

Step 4. Since we're not ready to teach graphics yet, SELECT **EXIT** from the Graphics Menu. This should take you back to the Main Menu.

Step 5. Some menu screens take you to sub-menus on the same screen. SELECT **SCREENS & REPORTS**. A sub-menu should appear on the right-hand side of the Main Menu screen (Figure 2-5). We really don't want to execute any of these functions; therefore, this is an ideal time to learn how to cancel a command. CANCEL the sub-menu. Remember, whenever we say CANCEL or you wish to cancel a command on your own, simply click the *right* mouse button. The sub-menu should have disappeared.

Before we move on to the final exercises for this lesson, we want to present alternatives to using the mouse. Instead of SELECTing via the mouse, you can do either of the following:

o   Use the space bar, tab key, or arrow keys to move through the menu options. Each time you press one of these keys, a different menu option will be highlighted on the screen. As soon as the desired option is highlighted, press the Enter key.

o   In every EXCELERATOR menu, each menu option begins with a unique letter or is assigned a unique number. Thus, the options can be selected by pressing the key for that letter or number.

Additionally, the Esc (escape) key on the keyboard is equivalent to the CANCEL command for the mouse.[3]

Exercise 2.3   Set Your System Profile Options

Whenever you log on to EXCELERATOR we recommend you initially review the latest system profile defaults for your account (those that were in effect at the close of the last work session on this account). A fellow EXCELERATOR user may have changed the defaults.

Step 1. SELECT the **HOUSEKEEPING** facility. A sub-menu has appeared to the right of the Main Menu (Figure 2-6). SELECT **Profile**. Another menu of options appears on the right-hand side of the screen (Figure 2-7). This menu is called an *Action Keypad* since the boxes correspond to actions that can be performed within the menu option you've selected. Action Keypads are used in several facilities.

Step 2. SELECT **Modify**. A screen similar to Figure 2-8 appears. This screen presents the default system profile settings for this account.

Step 3. If you have more than one printer in your system, you must decide which printer you wish to use. They are listed next to the heading HC DEVICE. If there is more than one printer, the default printer is highlighted with a dark background. To change the default, SELECT the alternate device. For practice, if you have two printers in the list, SELECT the other printer. Then, if necessary, SELECT the desired printer.

---

2   Throughout this tutorial, we will capitalize the words SELECT and CANCEL as they apply to the mouse functions. Furthermore, we will boldface the objects of mouse SELECTion actions (e.g., menu options, submenus options, commands, parameters, and the like.). We will also italicize EXCELERATOR messages and prompts, as well as anything that you should type.

3   Throughout this tutorial, we will only give instructions that correspond to the Microsoft Mouse input technique. Since the mouse must be used for several facilities, we suggest that you use the mouse for most or all facilities (except where indicated).

Figure 2-4

EXCELERATOR Graphics Menu Screen

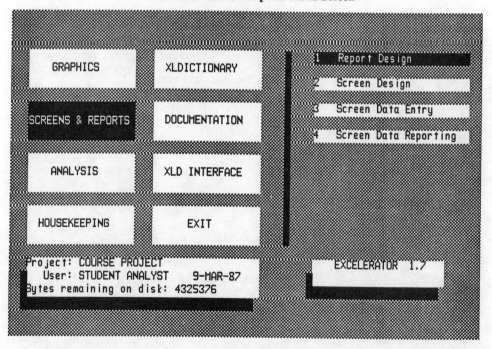

Figure 2-5

Submenu for SCREENS & REPORTS Facility

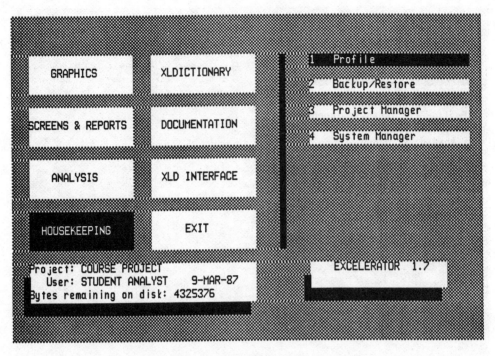

Figure 2-6

Submenu for the HOUSEKEEPING Facility

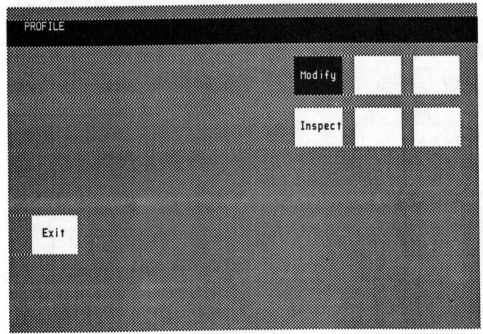

Figure 2-7

Action Keypad for Profile

20

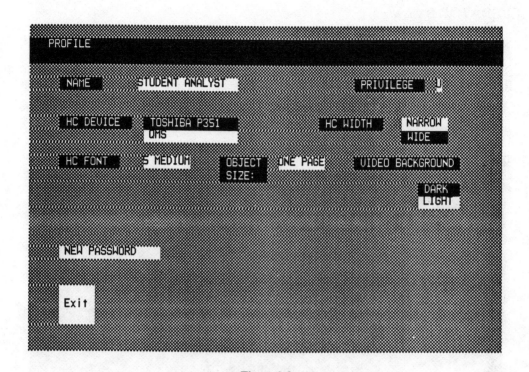

Figure 2-8

System Profile Attributes Screen

Step 4.   Most of our students prefer the **DARK VIDEO BACKGROUND** setting because it is easier on the eyes, especially when drawing graphs. If the default setting is not **DARK**, SELECT **DARK**.

Step 5.   For purposes of a feature we plan to demonstrate in Lesson 3 (and recommend for all EXCELERATOR graphs), SELECT HC WIDTH (which stands for HARD COPY WIDTH) **NARROW**. This orients the page such that line printing occurs across the 8.5 inch side of an 8.5 X 11 inch page. **WIDE** orients the page such that line printing occurs across the 11-inch side of an 8.5 X 11-inch page. The net effect of **NARROW** is to produce graphs that are oriented the same as the text reports they accompany. It is, admittedly, our preference.

Step 6.   Select **EXIT** to return to the Main Menu.

Exercise 2.4   Get Help From Within EXCELERATOR

From within EXCELERATOR's many functions, the F2 function key causes the system to display a one-line help message. Try it from the Main Menu. Use the Space bar or arrow keys (*not* the mouse) to highlight different Main Menu options. Then press the F2 function key. CANCEL (either from the mouse or by pressing the Esc key) removes the help message.

Exercise 2.5   Exit EXCELERATOR

Exiting EXCELERATOR is easy.

Step 1.   SELECT **EXIT**. A sub-menu appears to the right of the Main Menu (Figure 2-9).

Step 2.   SELECT **Leave Excelerator** from the sub-menu. This should return you to one of the following DOS prompts:

C>

C:\EXCEL>

If you want to quit this work session, this is an ideal time. Otherwise, proceed to Lesson 3.

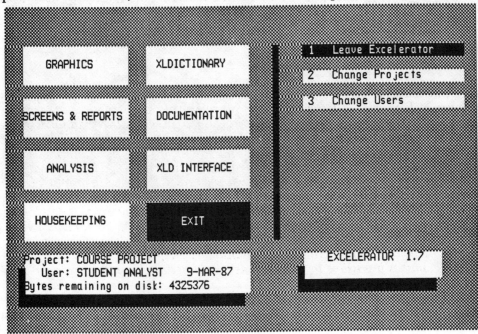

Figure 2-9

EXCELERATOR Exit Screen

Conclusion

In this lesson, you learned how to log on to EXCELERATOR. As a general rule, we recommend that you place the keyboard into Caps Lock mode since EXCELERATOR is case sensitive. We also recommend that you initiate every EXCELERATOR session by SELECTing the **HOUSEKEEPING**, **Profile**, and **Modify** actions, in that sequence. Review the subsequent default settings and change them as appropriate.

The use of a mouse is required in several facilities; therefore, we encourage you to use it in all facilities. If you are unfamiliar with the mouse, it may seem awkward at first. With experience, however, it will become second nature.

Now let's do some real work on the record club project. The next lesson introduces you to the graphics and dictionary features of EXCELERATOR.

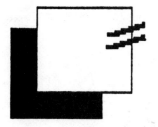

Part Two: Implementing Systems Development Methodologies

Systems development methodologies (SDMs) are specific tool/strategy combinations for implementing portions of the systems development life cycle. Although they are frequently sold as alternatives to the classical life cycle, they rarely accommodate all phases and situations encountered in projects. Examples of SDMs include Structured Systems Analysis and Design (Yourdon and IST), Information Engineering (Martin), Jackson Systems Development, Structured Requirements Definition (Orr), and prototyping (various software vendors). Note that we are excluding commercial project management methodologies, such as METHOD 1 and PRIDE, from our definition of systems development methodologies (although such methodologies appropriately acknowledge the sanctity of the life cycle and can use elements of SDMs).

The trouble with SDMs is that they usually impose greater rigor on the systems development process, so much rigor that many organizations have found them cumbersome. Frequently, documentation consists of many interrelated graphs and dictionary-like detailed specifications. It is easy to lose control of specifications -- a change to one specification can affect numerous other graphs and specifications.

EXCELERATOR makes it possible to get a handle on methodologies and specifications. The lessons in Part Two will teach you how to create specifications for the following popular methodologies:

o   data flow methodologies such as Structured Analysis
o   data modeling methodologies such as Information Engineering
o   rapid prototyping methodologies

We could not cover all methodologies. But these three have the greatest number of advocates. We are not endorsing a single methodology. We don't believe that any methodology encompasses all problems or the entire range of possibilities included in a well-conceived systems development life cycle.

First complete Lesson 3, an overview of GRAPHICS that is essential to all GRAPHICS options in EXCELERATOR. Lessons 4 through 6 cover data flow modeling, data modeling, and rapid prototyping, respectively. You can cover any or all of Lessons 4 through 6, in any order; depending on which methodology or combination of methodologies you use. However, to get full utility and value in Part Three of the tutorial, you should do all of the lessons. All of these lessons focus on the fundamental documentation aspects of the methodology. Part Three will look at advanced features and facilities that go beyond documentation.

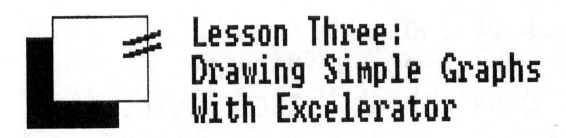

Lesson Three: Drawing Simple Graphs With Excelerator

The Demonstration Scenario

Let's begin the record club case study. The application we will design for Soundstage is a CUSTOMER SERVICES SYSTEM. That system consists of three key subsystems: subscription processing, monthly promotions, and order processing. They are described as follows:

    o   *The subscription subsystem handles new subscriptions, either through advertisements or referrals. This process normally results in new members and release of introductory offers or orders for those members.*

    o   *The promotion subsystem generates monthly and special offers to club members (in the form of promotions). Record-of-the-month promotions are released to all members. Members must respond to these promotions since they are dated and will automatically generate orders after a specified date.*

    o   *The order processing subsystem handles member orders and bonuses (free or reduced cost merchandise offered as an incentive for attracting new members or fulfilling membership requirements).*

In most projects an analyst studies documents, forms, and files; surveys users and managers; interviews users and managers; and the like to collect facts about problems and requirements. Most analysts then find it particularly useful to draw pictures of systems, either existing (during systems analysis) or proposed (during systems design). Historically, analysts have relied on paper, pencil, flowchart templates, and patience to draw these pictures. The patience stems from having to draw several drafts of diagrams and then redrawing them as mistakes and omissions are uncovered by users. In addition to the pictures, we catalog definitions, descriptions, and the like into a complementary glossary or dictionary.

The situation is as follows. You have just interviewed the Subscription Services Manager. He has described, in some detail, how new members' subscriptions are handled. Now you want to draw a picture that reflects the flow of data through his system. To make that picture easier for him to verify, you plan to describe physical aspects of the current system right on the picture.

What Will You Learn in this Lesson?

Much of EXCELERATOR's documentation potential can be directly accessed via a single facility, **GRAPHICS**. This potential includes access to the data dictionary. Graphics allow systems analysts to visually model systems from a variety of perspectives that can be more easily communicated to users and management. Graphics also support iterative analysis and design. In other words, graphs can easily evolve from depicting the current system to depicting the target system, including various alternatives generated along the way. You can easily experiment with graphs to modify designs or present alternatives. Finally, because objects on any graph can be described to the data dictionary, you can easily record and link to details that are not depicted on the graphs.

In this lesson you will learn how to use some of the common, basic commands associated with the **EXCELERATOR GRAPHICS** facility. You will also learn how to record objects from a graph into the **XLDICTIONARY**. After completing this lesson you will be able to:

1. Find your way around a typical **GRAPHICS** drawing screen including the *drawing area*, *drawing commands*, *orientation map*, and *status line*.
2. Use the **PROFILE** and **PRINT** drawing commands to change several of EXCELERATOR's default drawing options.
3. Use the **OBJECT, CONNECT, MOVE, DELETE, COPY, ZOOM,** and **DESCRIBE** drawing commands to draw a simple graph.
4. Use the **OTHER** or **EXIT** drawing command periodically to save your work.
5. Use EXCELERATOR's **HOUSEKEEPING** option to **Backup** your work on to floppy diskettes.

EXCELERATOR allows you to draw a variety of graphs including *data flow diagrams, structure charts, data model diagrams, entity-relationship diagrams, structure diagrams,* and *presentation graphs*. Since most people are familiar with data flow diagrams, we have chosen to draw a data flow diagram in this first lesson. The basic screen layout, commands, and procedures for drawing the other graphs supported by EXCELERATOR are similar.

The work that you do in this lesson will also be used in later exercises and lessons. In this exercise we will explain how to backup your work. Therefore, it is important that before proceeding with this exercise you have a formatted floppy diskette (you may need several before you complete the entire book).

Exercise 3.1   Draw Objects and Connections on a Graph

Let's learn how to draw a simple graph. Follow the steps described below. It is important that you follow the steps in the appropriate order.

> *Reminder*: When executing commands that involve multiple steps or SELECTions with the mouse, you can CANCEL the command by clicking the right mouse button. This is helpful when you realize that you've made a mistake.

Step 1.   Log on to EXCELERATOR and get to the Main Menu. Use **HOUSEKEEPING** and **Profile** to set default options (if you don't know how to do this, review or repeat Lesson 2).

Step 2.   SELECT **GRAPHICS** from the Main Menu. (Again, if you don't know how to SELECT menu options or use the mouse, review or repeat Lesson 2.) The Graphics Menu (Figure 3-1) has appeared. Notice that you have several graphics options when using EXCELERATOR.

Step 3.   SELECT **Data Flow Diagram** from the Graphics Menu. An Action Keypad appears in the upper-right quadrant of the screen (Figure 3-2). Since this is our first DFD, SELECT **Add**. You will be prompted for a name for your DFD. Naming conventions are *not* limited to 8 characters as they are in many microcomputer operating systems and programs. We can give our graphs much more descriptive names. For this DFD, type the name *SUBSCRIPTION PROCESSING*.[1] Press the Enter key.

---

[1]   If you are familiar with DOS file naming conventions, ignore them when using EXCELERATOR. In particular, do not give extensions, such as .DFD, to your names. EXCELERATOR automatically creates a DOS filename.DFD for each data flow diagram graph.

Figure 3-1

EXCELERATOR Graphics Menu Screen

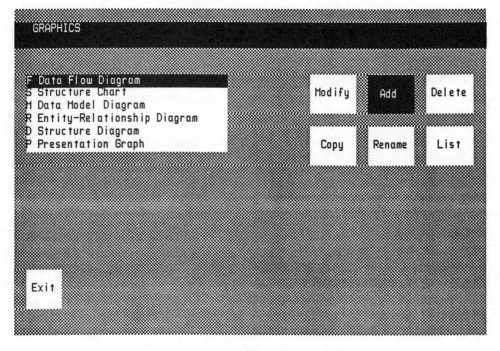

Figure 3-2

Action Keypad for Graphics Menu Screen

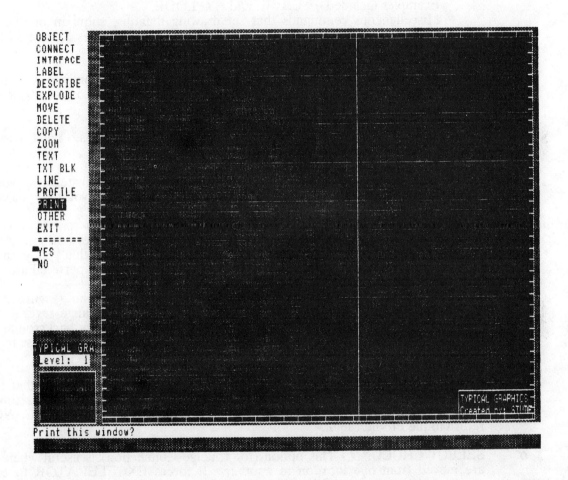

Figure 3-3

EXCELERATOR Graphics Drawing Screen

Step 4.   You should now see a typical Graphics *drawing screen* (Figure 3-3) for your data flow diagram. The areas of the drawing screen are described as follows:

o   Drawing area. This is where you will draw your DFD. The drawing area does *not* correspond to an actual 8.5 X 11-inch sheet of paper (it may be larger, depending of font size and object size settings established by your instructor or system manager).

o   Graphics commands. These commands create, delete, and move objects, create and modify object labels (names), initiate object entries into the data dictionary, and so forth. They fall into three types:

*   Drawing commands such as OBJECT, CONNECT, COPY, MOVE, DELETE, INTRFACE, LABEL, and TEXT.
*   Link commands that link the graphs and objects to the data dictionary details. Examples include DESCRIBE and EXPLODE.
*   Miscellaneous commands that set drawing defaults, zoom in on the screen, modify the display, and print individual graphs. Examples include PROFILE, PRINT, OTHER, and ZOOM.

As you'll soon see, some of the commands trigger additional commands that will be displayed in the same area of the screen (either below the main commands or in place of the main commands).

o   Orientation map. This is a miniature version of the drawing area. You'll learn more about it later.

o   Status line. EXCELERATOR uses this line to communicate with you. At appropriate times, the status line will prompt you for input, tell you that you've made errors, and display help messages.

Step 5.   Before you draw a DFD, we suggest that you set some default graphics options. Although your tastes may differ from ours, we suggest that you try our options so that you learn how to set options. Our options are based on extensive teaching experience and personal use.

SELECT **PROFILE** from the drawing commands. A different menu (Figure 3-4) has replaced the drawing commands. With this Profile Menu you can change several drawing options. The current settings for the options are highlighted with a dark background. After considerable experience with EXCELERATOR, we suggest that you use the mouse to change several of the settings. Proceed as follows:

o   SELECT **GRID**. An options menu appears. SELECT **FINE**. Notice that the tick marks around the drawing area are closer than they were. The fine grid setting gives you greater control in placing objects in the drawing area of the screen. CANCEL to return to the Profile Menu.

o   SELECT **CHGCONN**. This option (for Change Connection) is useful when objects are moved from one location to another. It forces EXCELERATOR to consider changing the side of the object which data flows (or other connections) enter or exit. Otherwise, the default setting will preserve the connection sides when you move objects.

o   CANCEL to exit the Profile Menu. This takes you back to the drawing commands menu.

You should always set your **PROFILE** options (as above) prior to drawing any new diagram. When you save a diagram only, the **PROFILE** options for that diagram are also saved. In time, you will develop your own preferences for the **PROFILE** settings. For more information consult your *EXCELERATOR Reference Guide*.

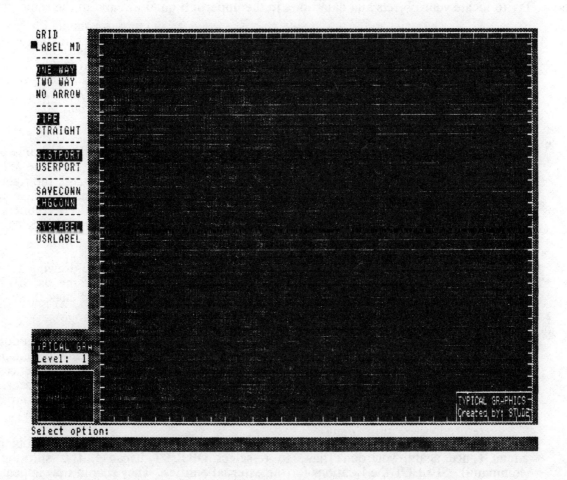

Figure 3-4

PROFILE Menu for Graphics

Step 6.   Before you draw the graph, there is one more thing you may want to do. Recall that the drawing area does *not* correspond to an 8.5 X 11-inch sheet of paper. How do you know precisely what portions of the diagram would be printed?

Here's a little trick we use. Before drawing the diagram, we *bound* it. SELECT the **PRINT** command from the drawing commands. A sub-menu appears. Now SELECT **FULL GRAPH**. Notice that the drawing area has been divided into four quadrants (Figure 3-5). The upper-left quadrant corresponds to a portion of the drawing area that would be printed on a single sheet of paper. Any portion of your diagram that crosses the boundaries of this quadrant will be printed on a separate sheet of paper.[2] Since we don't really have anything to print, CANCEL to escape the **PRINT** command. Notice that your quadrants are still being displayed!

You are now ready to draw your first diagram. We want you to reproduce the DFD depicted in Figure 3-6. Here's how. Try to locate your objects and data flows in the upper-left quadrant and in the same proximity as ours.

We should point out that EXCELERATOR supports two different sets of data flow diagram symbols. Figure 3-6 represents the Gane & Sarson symbol set and is the default. However, you're instructor or system/project manager may have changed EXCELERATOR to use the Yourdon & DeMarco symbol set (Figure 3-7). The two sets are equivalent. The remainder of the figures in this book were produced with the default setting (Figure 3-6).

Step 7.   SELECT **OBJECT** from the drawing commands. This SELECTion works the same as other menus. Next, use the mouse to move the cursor to a command and click the *left* mouse button. If you choose the wrong command or want to change the command, the right mouse button CANCELs the current command. If the system beeps, you've made a mistake. The Status Line should clarify the problem. If you correctly selected the **OBJECT** command, a list of the valid object types will appear in a separate menu below main graphics commands (Figure 3-8).

Step 8.   SELECT **PROCESS** from the object types. SELECT the location in the drawing area where you want the process to appear. SELECTing locations works the same as SELECTing commands and menu options. Just move the cursor to the location and click the *left* mouse button. The object has appeared.

Even though the Status Line says *Select object type*, there is actually no need. You don't have to re-select either **OBJECT** or **PROCESS** because they are still the *active* commands. By *active* we mean that EXCELERATOR is still expecting you to draw new occurrences of the same object type. SELECT another location where you want a process to appear. Now SELECT the location of the third process. You should see three processes in your drawing area.

Step 9.   We also want to draw some external entities. SELECT **X ENTITY** from the object types menu (once again, you don't have to re-select **OBJECT** because it is still the active command). SELECT the locations for the external entities. They should now appear on the screen.

Step 10.   Now we can add the two data stores. SELECT **STORE** from the object types sub-menu (again, **OBJECT** is still the active command). SELECT the locations. All objects except for the data flows now appear on the screen.

Step 11.   Notice that the data flows are not included in the object type menu. Data flows are drawn using the **CONNECT** option from the drawing command menu. EXCELERATOR refers to

---

2   If the printer selection, object sizes, and font sizes are set up as recommended in the Guidelines for Setting Up the Tutorial Environment (preceding the Preface), the upper-left quadrant corresponds to an 8.5 X 11 inch page.

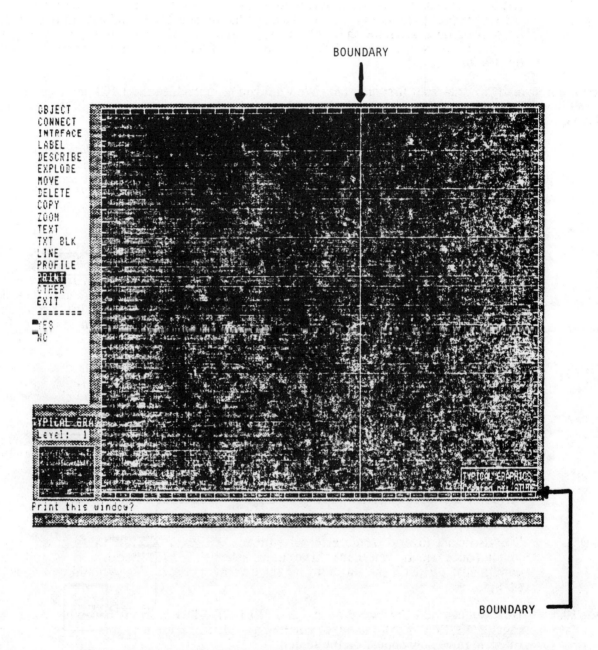

Figure 3-5

Printing Quadrants on the Graph

31

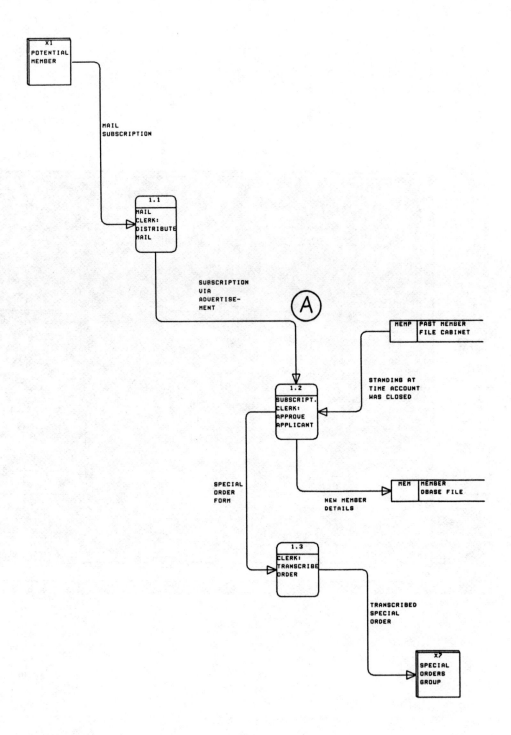

Figure 3-6

Subscription Processing Data Flow Diagram (Gane & Sarson Symbology)

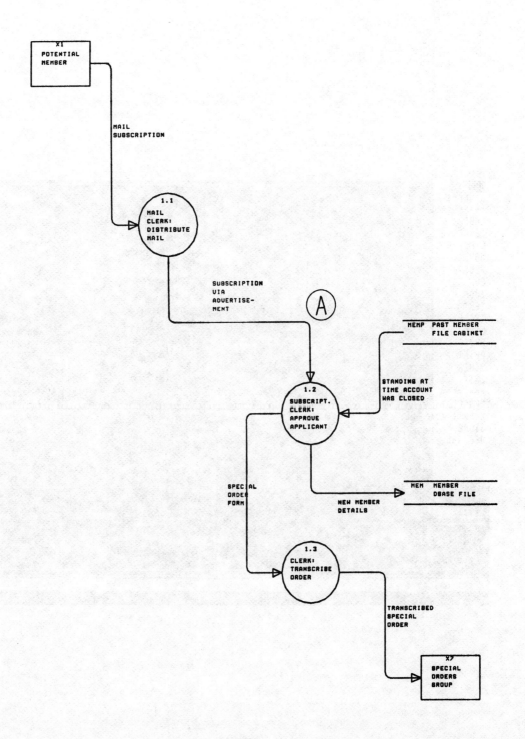

Figure 3-7

Subscription Processing Data Flow Diagram (Yourdon Symbology)

33

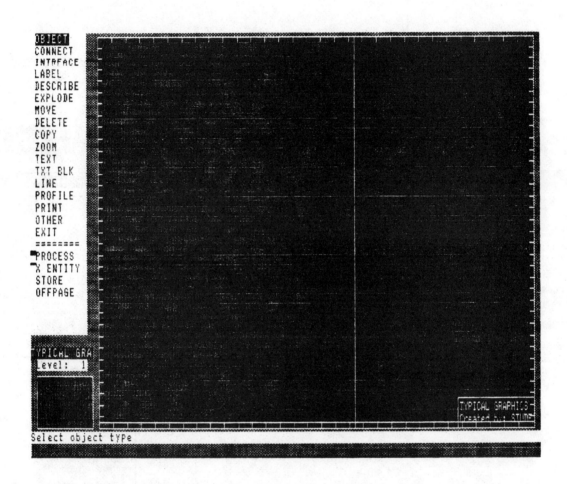

Figure 3-8

Object Types Submenu on Graphics Drawing Screen

34

the data flows as *connections*. Let's draw the MAIL SUBSCRIPTION data flow from Figure 3-6 or 3-7. SELECT **CONNECT**. Look at your status line. EXCELERATOR is instructing you to *Touch first object*. Respond to this request by SELECTing an object *from which* a data flow will originate. Look at the status line again. EXCELERATOR is now instructing you to *Touch a second obj or an intermediate location*. For this step, just SELECT the other object *to which* you want the data flow to go. The data flow should appear on your screen. Don't draw the other data flows yet.

Step 12.   Now let's draw the SUBSCRIPTION VIA ADVERTISEMENT data flow (still from Figure 3-6 or 3-7), but using a slightly different approach. **CONNECT** is still the active command so we don't need to re-select it (even though the Status Line is telling you to *Select Command*). Instead, we want you to SELECT the process from which SUBSCRIPTION VIA ADVERTISEMENT originates. Once again, the Status Line suggests that you *Touch a second obj or an intermediate location*. This time you'll SELECT a location instead of an object. Looking at Figure 3-6 or 3-7, touch the location where we've added the letter "A." Notice that a partial data flow has been drawn. Now respond to the Status Line by SELECTing the second process. The data flow is completed. This technique gives you somewhat greater control over the appearance of data flows. You could have SELECTed several locations to insert multiple curves in the data flow. In other words, you can build the connection in segments.

Step 13.   Now draw the two data flows to/from the data stores, repeating the technique taught in Step 11 (*not* Step 12). It doesn't matter which data flow you draw first. Notice that the data flow connections to the process are very close. Sometimes you may not agree with EXCELERATOR's choice of entry or exit *port* (a *port* is the exact location of the entry or exit). And sometimes EXCELERATOR will cross data flows (giving the picture a more complex, harder to read look). Fortunately, you can move data flows and change the port selections to give your diagram a cleaner appearance. Here's how:

o   **SELECT MOVE.**

o   You may have already noticed the little square boxes that overlap the data flows. The are called *handles*. They give you something to grab or touch when applying commands to data flows. Let's move the exit port for the data flow that connects the *MEMBER* data store to the process. SELECT the data flow by touching that data flow's handle.

o   Notice that a handle has now appeared at each end of the data flow you selected. Now SELECT the handle at the end that you want to change. Now SELECT the process to which the data flow is connected.

o   Notice that a small arrow appears along the outside of the process. This arrow can be moved along the outer border of the object using the mouse. Try it. Move the mouse across a surface and notice that the arrow moves accordingly. Also refer to the Status Line. The Status Line identifies the specific side and port location pointed to by the arrow. Move the small arrow to the desired side and port location for the connecting data flow and press the left mouse button. The data flow should redraw itself to the new location. With this technique you can change the entry and exit ports for any data flow.

o   Suppose you accidentally connect one end of a data flow to the wrong process. It's easy to fix. Just SELECT **MOVE**. Then SELECT the handle of the data flow. Handles will appear on both ends of that data flow. SELECT the end which is incorrect. SELECT the correct object (or a series of locations leading to the correct object). Finally, SELECT the desired side and port location. The data flow should be moved. Try it! Move a data flow connection to a different object. Then, move that data flow back to its original object. Easy!

35

Step 14.   Draw the remaining data flows using the above technique. Your diagram should look similar to Figure 3-6 or 3-7.

Step 15.   Before you name the objects and data flows, you really should learn a little more about the power of the **GRAPHICS** facilities. One of the classic problems with all graphics techniques is *modification*. Imagine that the user forgot to tell you something, you misunderstood something, you just came up with a better idea, or you just want to improve the graph's appearance.

You've already learned how to use the **MOVE** command to move data flow connections. **MOVE** is also used to move objects. SELECT **MOVE**. SELECT a process, data store, or external entity - any object. Now SELECT a new location for that object. The object and all connecting data flows are relocated. Practice this command with other objects and then get the DFD back to its original appearance (Figure 3-6 or 3-7).

In the lower right-hand quadrant of the drawing area, you should see a title block with the name of your DFD. **MOVE** that text block into the upper left-hand quadrant of the drawing area. We like to do this for all graphs.

Step 16.   You can also make exact copies of objects. For instance, you might want to copy an external entity or data store to make it possible to avoid crossing data flows on a DFD. Let's practice. SELECT **COPY**. SELECT an external entity of your choice. SELECT a location for the copy. Had the objects already been labeled, the labels would have also been copied. Duplicate external entities and data stores are sometimes used to improve the appearance of DFDs (by eliminating connections that cross one another).

Step 17.   You can also **DELETE** objects or data flows. The technique is similar to **COPY**. Let's delete the copy of the external entity you just added to the graph. SELECT **DELETE**. SELECT the object you want to delete. It's gone! **DELETE** also works with data flows. Be careful with **DELETE**. If you delete a data flow, only the data flow is deleted. However, if you delete an object, *all* of the data flows that were connected to that object are deleted.[3] If you want to practice, add some miscellaneous objects and data flows to your graph and then delete them.

Step 18.   When you are working with EXCELERATOR in long sessions, it is a good idea periodically to save your work. Let's practice. SELECT **OTHER**. A sub-menu appears. SELECT **SAVE**. Your work is now saved, and you can still continue work with your graph.

Step 19.   To save and leave the graph, SELECT **EXIT**. If you hadn't already saved your work, EXCELERATOR would have asked you to **SAVE** or **NO SAVE**. If it asks, you should always SELECT **SAVE**, just to be safe. You have now exited back to the screen that listed all of the graphs that EXCELERATOR can draw. SELECT **Exit** again to get back to the main menu.

Step 20.   If you want to quit this work session, you should first skip to Exercise 3.3 to learn how to **Backup** your work. That exercise will also teach you how to **Restore** your work in the next work session.

Exercise 3.2   Naming Objects and Connections on a Graph

Now we are ready to name the objects and connections appearing on the graph you created in Exercise 3.1. If you are not starting a new work session, go to Step 1, skipping the rest of this paragraph. Otherwise, if you are starting a new work session, you should log on, reset your **Profile**, and **Restore** your work (covered in

---

[3]   If the object has been described to the dictionary (covered later in this exercise), you would be prompted as to whether or not you want the dictionary description deleted also. In most cases, the answer is "no"; therefore, the default is n.

Exercise 3.3). Then, SELECT **GRAPHICS** from the Main Menu, SELECT **Data Flow Diagram** from the Graphics Menu, SELECT **Modify** from the Action Keypad, type *SUBSCRIPTION PROCESSING* for the name, and press the Enter key.

Step 1. The objects are too small to be easily labeled and read. Fortunately, we can zoom in on portions of our graph. **SELECT ZOOM.** The **ZOOM** command will change the magnification of objects on the screen. A sub-menu appears below the drawing commands. You are currently in the default mode, **LAYOUT**, which shows the entire graph on the screen. **SELECT CLOSE UP.** Now we have to tell EXCELERATOR what portion of the graph to zoom in on. The Status Line says, *Touch location in orientation map*. Recall that the Orientation Map is in the lower, left-hand corner of the screen.

Familiarize yourself with the affect of **ZOOM**ing by SELECTing several locations on the orientation map. Notice that a ZOOMed-in portion of the graph appears in the drawing area. SELECTing several locations on the orientation map is called *panning the graph*. The little dots on the map correspond to object locations in the drawing area. Conclude this step by panning to the location of the POTENTIAL MEMBER external entity in the upper left-hand corner. The external entity object should appear within the drawing area.

Before you begin naming the objects and connections appearing on your graph, you should understand that there are two types of names that can be assigned to an object or connection. The terminology can be a source of confusion; therefore, we will try to differentiate. The terms are described as follows:

o The terms *Name* and *ID* are both used to identify and describe graph objects to the data dictionary. In the **GRAPHICS** facility, the term *ID* is used. In other facilities, the term *Name* is used. IDs or names are up to 32 characters in length.

o The term *Label* is used to describe the name that appears inside an object on a graph. Depending on the settings for printer, object size, and font size, *labels* can either be larger or smaller than *IDs*.

Obviously, we want to *label* the objects and connections on our graph. There are three ways to label objects. One way is dead wrong and will get you into trouble. Another technique works but can get you into trouble. We'll begin with the recommended technique (command) and then briefly explain the other two commands:

o **DESCRIBE.** We recommend that **DESCRIBE** *always* be used for naming and labeling all objects and data flow connections. **DESCRIBE** accomplishes two things. First, **DESCRIBE** allows you to give the object or data flow an ID *and* a label. Second, and at the same time, **DESCRIBE** initiates an entry for the object or data flow into your data dictionary. Once in the dictionary, the description and label is available for reuse and analysis (covered in other lessons).

o **LABEL.** This command can legitimately be used to assign a label to your objects and data flows. However, we emphatically recommend that you *not* use **LABEL** to label any object. Why? **LABEL** does not enter your object into the data dictionary, that's why! You may feel that, "I can always add my objects to the dictionary later." This is true; however, our experience shows that as the number of objects and details associated with the project grows, there is a tendency to forget to enter the objects into the dictionary. Besides, why take two steps to do what you can do in one step using the **DESCRIBE** command. Until you've **DESCRIBE**d objects to the dictionary, they are only images on your graph. Furthermore, objects must be **DESCRIBE**d to the dictionary to exploit EXCELERATOR's full analytical and maintenance capabilities.

o **TEXT** and **TXT BLK.** These commands were provided by EXCELERATOR so that you can add a line of text (**TEXT**) or a block of text (**TXT BLK**) at any location in the drawing area. Unfortunately, some people use **TXT BLK** to name processes, external entities, data stores, and data flows (using text that overlaps the object). This is wrong! If you use **TXT BLK** to name objects and data flows, not only do you eliminate the possibility of including the name in your data dictionary, but you also make it impossible to **MOVE** objects and data flows to different locations in the drawing area (the text block doesn't move with the object). Only use **TEXT** and **TXT BLK** to add miscellaneous comments to your diagrams.

Let's now label our objects.

Step 2.   SELECT **DESCRIBE** from the Drawing Commands Menu. Then SELECT the top-most external entity from the drawing area. The Status Line now instructs you to *Enter ID*. The ID will serve as an index or key that references the object. IDs are a powerful concept. In later lessons, you will see the importance of being able to use the ID to reference objects and data flows. Specific ID naming recommendations will be made in Part Three of the tutorial. For the time being, give this object the simple ID of *X1*. Press the Enter key.

Step 3.   You should now see an External Entity Description Screen (Figure 3-9). For the first time, you are into the data dictionary facility of EXCELERATOR! The **Label** box on the screen is used to add a label to an graph object. The cursor is already positioned in the **Label** box.

The **Label** block does support *character wrap* but does not support the *word wrap* feature associated with a word processor. Therefore, *you* must ensure that words do not get split over multiple lines. You do this by adding spaces as needed. Start a new line by pressing the Enter key. You can also move around the block with the arrow keys on the numeric keypad (right-hand side of most keyboards).

On the first line of the **Label** block, type *POTENTIAL*. Press the Enter key. On the second line, type *MEMBER*. We suggest that you minimize the use of abbreviations on **Labels**.

Step 4.   Now use the mouse to SELECT the first position in the **Description** block. The **Description** block is a free-format text block. It is actually a scrolling block that allows you to enter large amounts of additional detail (up to 60 lines of 72 characters each) about the object. EXCELERATOR's **Description** text editor is similar to a word processor in that it supports word wrap. In other words, you don't have to press the Enter key to go to a new line since any full word that doesn't fit on a line will automatically wrap around to the next line as you type. You can, of course, force a new line by pressing the Enter key (the paragraph symbols that appear on the screen won't print)[4]. Additionally, the text editor provides a number of cursor and editing keystrokes as defined in Table 3-1.

For practice, type the following paragraph (without ever pressing the Enter key): *A POTENTIAL MEMBER IS A PERSON WHO IS EITHER RESPONDING TO ONE OF THE RECORD CLUB'S ADVERTISEMENTS OR WHO HAS BEEN RECRUITED OR REFERRED BY AN EXISTING MEMBER.*

To exit any data dictionary screen, press the F3 function key. F3 means "Save and exit the dictionary."[4] This returns you to your original facility, in our case, the data flow diagram we've been working on.

Step 5.   Now let's label a data flow. **DESCRIBE** is still the active command so we don't need to SELECT it even though the Status Line says so. SELECT the data flow *MAIL SUBSCRIPTION* (from Figure 3-6 or 3-7). You do this by SELECTing the data flow's handle. Once again, you are asked for an ID in the Status Line. Type *ORD: MAIL SUBC* and press the Enter key. Once again, your in the dictionary. The Data Flow Description Screen looks a little different (Figure 3-10); however, the cursor is still in the **Label** block. Type the label *MAIL SUBSCRIPTION* and press the F3 function key. You have returned to the graph and the label appears next to the data flow. The IDs for data flows are not displayed. (With the default options, both IDs and labels are displayed for all other objects.)

---

4   If you are using Version 1.6 or earlier, you must press the Tab key to force a new line break. Use of the Enter key in Version 1.6 or earlier will send you back to the graph. Also, earlier versions did not support word wrap.

Figure 3-9

External Entity Description Screen (in the Data Dictionary)

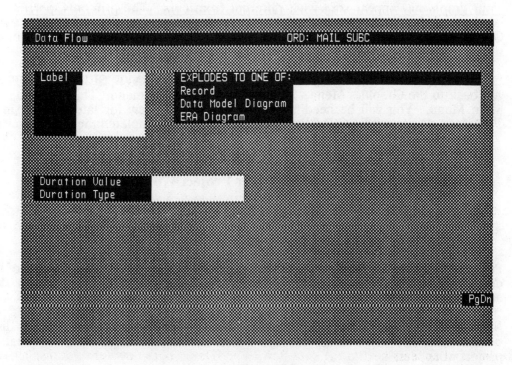

Figure 3-10

Data Flow Description Screen (in the Data Dictionary)

Step 6.   Now, practice using **DESCRIBE** to label all of the remaining objects and data flows (as in Figure 3-6 or 3-7). Make sure you use the IDs that we used in our figure. For data flows, use abbreviations of the labels for the IDs (you'll learn why in the next lesson).

Invariably, the question arises, "Can I move my data flow labels to alternate locations?" EXCELERATOR doesn't always place the labels in the best location. Labels may overlap. Or you might find it difficult to tell which label belongs to which data flow.

Step 7.   Pick a data flow label that you'd like to move and determine where you want to move it. **SELECT MOVE.** Did you notice that brackets have formed around every visible data flow label on the screen? SELECT any position inside the brackets that enclose the label you want to move.

SELECT the location where you would suggest that the data flow label be moved. The label has now moved to that location.

Labels cannot be deleted with the **DELETE** command. You can delete a label by typing spaces over the existing label.   You've now finished your first EXCELERATOR graph.

Step 8.   Suppose you want to print your DFD. We recommend that you always save your work before printing. **SELECT OTHER. SELECT SAVE.** Within moments, your work is saved.

Now **SELECT PRINT.** You're now prompted for the appropriate graph size. **SELECT FULL GRAPH.** The full graph reappears. So long as you have no text, connections, or objects outside of the upper right-hand quadrant, your graph will print in one page (a good idea!). EXCELERATOR now asks you to confirm printing of the graph size. **SELECT YES** to confirm that you want to print the graph. You're now asked whether the output is to be printed to a file or printer. **SELECT printer.** Your graph should momentarily be printed.

Your graph may appear somewhat different from ours. The printouts, percentage of each labels printed, and the like, are all functions of your printer, the object size, and font size. Our graphs were printed on a QMS LaserGraphix, one of the highest resolution printers supported by EXCELERATOR.

Step 9.   Let's exit. **SELECT EXIT.** If prompted to **SAVE** or **NO SAVE, SELECT SAVE.** This takes you back to the Graphics Menu. SELECT **Exit** from that menu. This takes you back to the Main Menu. You will be needing the graph you just drew for later lessons in this book. Therefore, do *not* exit EXCELERATOR until you have completed the next exercise.

Exercise 3.3   Backing Up and Restoring Your Project Work

Experienced computer users frequently backup their work. With EXCELERATOR, we recommend backups at the end of every work session for the following reasons:

o   Project directories for projects tend to get quite large. Thus, the consequences of losing one's data can be severe -- the loss of work for several days, weeks, or months. Hard disks do crash, and if they do, the data on those disks may be permanently lost.

o   Many EXCELERATOR workbenches are shared by several analysts. Hard disk capacity becomes the key constraint. Backups guard against accidental or intentional erasure of project work files.

o   Many EXCELERATOR project accounts are shared by several analysts working on the same team. If the hard disk on the workstation contains several other software packages, several teams may also have to share accounts. With shared accounts and project directories, the potential for erasure and errors increases.

When using EXCELERATOR, we highly recommend a father/grandfather backup scheme. This requires two separate sets of backup diskettes, labeled *BACKUP A* and *BACKUP B*. The most recent backup diskettes are always referred to as the *father*. The oldest backup diskettes are called the *grandfather*. When backing up, you always use the grandfather diskettes (BACKUP A or BACKUP B). After the backup, that set always becomes the father. The other set (previously the father) becomes the grandfather. Using this method, you never can lose more than two work sessions of work (usually, the father set ensures that you will lose no more than one session of work). Data processing professionals are very familiar with this technique. Students may be less familiar. It would probably be a good idea to keep a log so you always know which backup set, A or B, is the father and which is the grandfather.

In order to complete the exercise, you'll need to have at least one *formatted* diskette. Consult your instructor or system or project manager to learn how to format diskettes. Diskettes should be numbered sequentially and used in that order (never label the diskette with a pencil or ball point pen since you could damage the diskette with pressure). Given formatted, labeled diskettes, backup is performed as follows:

Step 1. From the Main Menu, SELECT **HOUSEKEEPING**. From the sub-menu that appears (Figure 3-11), SELECT **Backup/Restore**. From the next screen, SELECT **Backup**.

Step 2. The screen adds prompts for **DATE:** and **PROJECT NAME:**. We have found that it is best to *not* enter a date. Simply press the Enter key to move on to the **PROJECT NAME:** prompt. Type the project name and press the Enter key.

Step 3. The system has temporarily transferred control from the EXCELERATOR software to DOS's **BACKUP** command. The system will prompt you to insert Diskette 1 into the drive and press Enter. **CAREFUL!** DOS's backup erases all files from the root directory on the diskette (usually the entire diskette). Don't use a diskette containing files you can't afford to lose.

DOS will prompt you to insert new diskettes as needed. It is important to insert the diskettes in the same order as they are numbered.

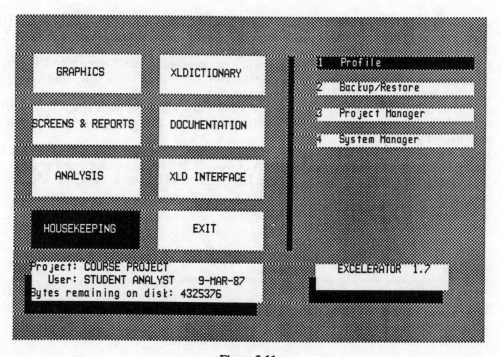

Figure 3-11

HOUSEKEEPING Submenu Screen

Step 4. When the backup is complete, you are returned to EXCELERATOR's **Backup/Restore** screen. SELECT **Exit**. This returns you to the Main Menu. It is now safe to exit EXCELERATOR. SELECT **EXIT**. Select **Leave Excelerator**.

When you begin your next work session, you should restore from your father backup. For practice, let's do it.

Step 5. Log on to EXCELERATOR and get to the Main Menu.

Step 6. SELECT **Backup/Restore** from the **HOUSEKEEPING** sub-menu. SELECT **Restore**. You will be prompted for a **PROJECT NAME:**. If the project name is the one you selected when logging on, press the Enter key. Otherwise, type the project name and then press the Enter key.

Step 7. You have temporarily exited EXCELERATOR and entered DOS's **RESTORE** command. The system will prompt you to insert Diskette 1 into the drive and press enter. Always use your father backup diskettes unless they are bad (in that event go to your grandfather diskettes). You must insert the diskettes in the same sequence as the backup (which is why you labeled them 1, 2, 3, and so forth.

Step 8. When the restore is complete, you are returned to EXCELERATOR's **Backup/Restore** screen. SELECT **Exit**. This returns you to the Main Menu. Now you can begin your work. However, this step completes this lesson. SELECT **EXIT**. SELECT **Leave Excelerator**.

Now that you know how to backup your work and restore it, you can safely end work sessions at any time in this tutorial.

Conclusion

Drawing systems analysis and design graphs is easy in EXCELERATOR. From the **GRAPHICS** facility you can draw several types of graphs. Graphs are drawn on a drawing screen. Using the mouse, you SELECT commands and locations to create, delete, and modify objects and connections. Objects should always be labeled using the **DESCRIBE** command - even though a separate **LABEL** command exists. This way, all objects are cataloged into the data dictionary. Cataloging objects into the dictionary is important if you want to take advantage of EXCELERATOR's full potential.

You also learned how to backup and restore your work. Backup and restore will protect you from accidental loss of data. These commands also make it possible for you to move through the remaining lessons and exercises at your own pace.

Most users of EXCELERATOR use variations on formal systems development methodologies (for instance, *Structured Systems Analysis*). EXCELERATOR is fully capable of supporting such methodologies. In the next three lessons you will learn how to use EXCELERATOR to create specifications according to three of the most popular methodologies.

Lesson Four:
Structured Analysis and
Data Flow Modeling

The Demonstration Scenario

Soundstage Record Club's Data Processing Department is predominantly a COBOL shop. They also use IDMS, a Database Management System. The Systems Department has adopted a methodology that represents their interpretation of three popular methodologies:

    o   *Structured Systems Analysis and Design (for all COBOL applications)*
    o   *Information Engineering (to assist with database aspects of all applications)*
    o   *Prototyping (for requirements determination and design of all user interfaces such as reports and screens)*

Structured Systems Analysis and Design is based on the flow of data through a system. Analysts study that data flow for the current system and design the data flow for an improved system. Detailed data dictionary specifications and logic specifications are developed as the project progresses.

    It is relatively early in the project. You and your fellow analysts have interviewed all users and managers in the Customer Services Division and its constituent departments. You are in the process of drawing data flow diagrams, the pivotal tool of Structured Analysis, for the entire system. Specifically, you are drawing "physical" data flow diagrams to confirm your understanding of both the data flow, but also the media for data flows, who or what processes those flows, and where data is stored in the current system. Experienced structured analysts would eventually transform these physical DFDs into "logical" DFDs that represent "what" the system does without the constraint of "how" the current system accomplishes those tasks.

What Will You Learn in this Lesson?

In this lesson you will learn how to use EXCELERATOR to support documentation for data flow and process modeling. Data flow and process modeling are techniques associated with the very popular *Structured Systems Analysis and Design* methodology, as advocated by DeMarco, Yourdon, Gane, and Sarson. System models reflect the flow of data through the system and the work performed on that data as it flows through the system. Details are revealed gradually since the methodology uses a top-down system decomposition strategy.

    The central tool in *Structured Systems Analysis and Design* is the **data flow diagram (DFD)**. The strategy begins with drawing high-level diagrams of the system followed by recursive decomposition of the higher-level diagrams to diagrams that reveal greater detail. Diagrams iteratively evolve from those which model the current system to those which model system alternatives and, finally, to those which model the new system. Ultimately, the data flows, data stores, and processes depicted on the the detailed DFDs are decomposed into detailed specifications. An entire *Structured Analysis* system specification will also include a comprehensive **data dictionary** and **mini-specs**, both of which can also be developed directly from EXCELERATOR's **GRAPHICS** facility. This lesson will show you how to create *Structured Systems Analysis* specifications using EXCELERATOR. Specifically, we'll focus on the documentation aspects of EXCELERATOR. You will know you have mastered this exercise when you can:

1. Use EXCELERATOR's **GRAPHICS** and **Data Flow Diagram** facility to plan or outline the overall structure of a system.

2. Use EXCELERATOR's **GRAPHICS** and **Data Flow Diagram** facility to draw a leveled (using the **EXPLOSION** command) set of data flow diagrams including:

   o   a context data flow diagram
   o   a system-level data flow diagram
   o   middle-level data flow diagrams
   o   primitive-level data flow diagrams

3. Use the ability of EXCELERATOR's **GRAPHICS** and **Data Flow Diagram** facility to decompose data flows, data stores, and processes into detailed specifications including:

   o   data structure as specified in record layouts
   o   data structure as specified in data models plus record layouts
   o   Structured English mini-specs

EXCELERATOR is also capable of checking the *Structured Analysis* specification for errors and inconsistencies; however, that facility will be deferred until Lesson 8. Lesson 4 builds on the basic **GRAPHICS** skills that you learned in Lesson 3.

As a reminder, EXCELERATOR supports both popular DFD symbol sets. They were demonstrated in the last lesson. The figures in this book were produced with the default setting, called the Gane & Sarson symbol set. Your diagrams may use the alternate symbols if your account has been so established. Also, this lesson is based on *physical data flow diagrams*.

Finally, as we start the lesson, we just want to remind you that it is not our intent to teach or advocate *Structured Systems Analysis and Design* in this tutorial. Some familiarity with the methodology or tools would be helpful. Also, be aware that every so-called structured analyst (or organization) has different standards or conventions. Consequently, our approach may not precisely match yours. However, the intent of the lesson is to introduce the power of EXCELERATOR as a support tool for the basic concepts of data flow and process modeling.

Exercise 4.1   Draw a Hierarchy Chart as an Outline for Data Flow Diagrams

As you learned in the last lesson, data flow diagrams (DFDs) are one of several graphics tools directly supported by EXCELERATOR. We acknowledge that true *Structured Analysis* traditionally begins with a context DFD. The analyst then proceeds to level the context diagram into a system diagram, middle-level diagrams, and, finally, primitive or detailed-level diagrams. Unfortunately, we have found that the structure and organization of these so-called leveled DFDs are frequently haphazard. We want to show you an outlining feature made possible through EXCELERATOR's data dictionary. This feature uses the power of the dictionary to help you plan or derive a more sensible top-down structure when using *Structured Analysis*. The final product also helps you to more quickly build the diagrams.

Good writers try to organize their thoughts before writing. The same principle can be applied to data flow diagraming. Instead of a classical outline, we will draw a hierarchy chart. Although hierarchy charts are typically thought of as structured design tools, they can be equally useful for planning the explosion levels for *Structured Analysis*. The hierarchy chart for our record club case study is presented in Figure 4-1. Your job is to recreate our hierarchy chart.

Step 1.   Log on to EXCELERATOR and set your **HOUSEKEEPING Profile** options. SELECT **GRAPHICS** from the Main Menu. SELECT **Data Flow Diagram** from the Graphics Menu. That's right! We're going to use the DFD facility to draw our hierarchy chart. You'll soon see

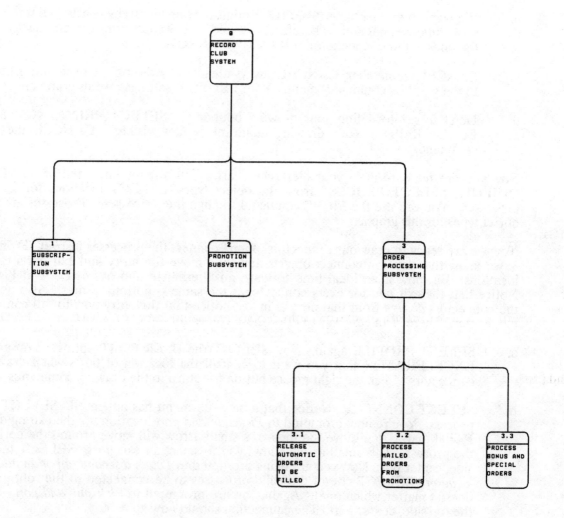

Figure 4-1

Hierarchy Chart Outline for Record Club Project

why. SELECT **Add**. For this graph, type the name *RECORD CLUB SYSTEM OVERVIEW*. Press the Enter key or F3 function key.

Step 2. You should now see the drawing screen. Next, set the graphics **PROFILE** options. You did this in Lesson 3. By way of review:

    a. SELECT **PROFILE** from the graphics commands. SELECT **GRID**. SELECT **FINE**. CANCEL to return to the **PROFILE** menu. SELECT **CHGCONN**.

    b. We want to set one new **PROFILE** option for our hierarchy charts. Hierarchy charts show *structure*, not *flow*. Therefore, to avoid reader confusion, let's not place arrows on the ends of our connections. SELECT **NO ARROW**.

    c. CANCEL (remember, CANCEL always means click the right mouse button) to return to the graphics commands menu. Your **PROFILE** settings for this graph are complete.

    d. Don't forget to define your drawing boundary. SELECT **PRINT**. Now SELECT **FULL GRAPH**. Your drawing quadrant is now visible. CANCEL the printing command.

Step 3. Now you are ready to draw your hierarchy chart. You already know the basics. SELECT **OBJECT**. SELECT **PROCESS** from the object types. SELECT locations for all of your processes. You can use the **MOVE** command to align the processes. Processes are the only object types for this graph.

Step 4. Now we are ready to draw our connectors. Don't connect the processes just yet. From Lesson 3, we know that you can connect objects and then move the entry and exit ports to desired locations. But now is an ideal time to learn an alternative approach. Look at Figure 4-1. Notice that the exit *port* for every connector is the same -- bottom, center. Also notice that the connectors *diverge* from that port. Finally, notice that the entry port for all connectors is top, center. There is an easy way to draw these connectors.

    a. SELECT **PROFILE** again. Now SELECT the **USER PORT** option. You are telling EXCELERATOR that you want to control the location of ports when drawing new connectors. Click the right mouse button to return to the drawing commands menu.

    b. SELECT **CONNECT**. Notice that a new sub-menu has appeared. SELECT the root process. You're now prompted to *Pick side and port* location for the exiting data flow. Recall that as you move the mouse, a small arrow will move around the perimeter of the process. This small arrow allows you to select an exit port. SELECT the bottom side, center port. Now you are being prompted to *Touch a second obj or an intermediate location*. SELECT one of the child processes to be connected to the root process (it doesn't matter which one). Again, you are prompted to *Pick side and port*. SELECT the top side, center port. The connection should now appear.

    c. Now add the remaining connections using the same technique. Be sure to use the exact same side and port locations to ensure connections will appear as diverging. Note that there is no need to re-SELECT the **CONNECT** command. The **CONNECT** command is still active. Simply re-SELECT the root process.

Step 5. Now you are ready to label the processes. Use the same IDs and labels as shown in Figure 4-1. Recall that the objects are too small to be easily read. SELECT **ZOOM**. SELECT **CLOSE UP** or **MEDIUM**. Pan to the location of the first process you want to name. Also, recall that we always want to use **DESCRIBE** (not **LABEL**) since it initiates a reusable entry for the object into your data dictionary. You'll soon need those entries.

The procedure for naming objects such that they are cataloged into your dictionary was taught in Lesson 3. By way of review, SELECT **DESCRIBE**. Then SELECT a process from the drawing area. Enter the ID (from Figure 4-1). Press the Enter key. Enter the **Label** on the data dictionary screen and press the F3 function key to return to the graph. **DESCRIBE** the remaining processes in the same manner.

Step 6.   That completes Exercise 4.1. **SELECT OTHER. SELECT SAVE.** You may want to **PRINT** a copy of your hierarchy chart before proceeding to the next exercise. Printing was covered in Lesson 3.

SELECT **EXIT** from the graphics commands. This takes you back to the Graphics Menu. SELECT **Exit** from the Graphics Menu. If you don't plan to proceed immediately, you might consider doing a **Backup** (via **HOUSEKEEPING**). Backup were covered in Lesson 3.

If you can't fit an entire hierarchy chart for a system into a single chart, **Add** additional charts, repeating appropriate root processes on the subsequent charts.

Exercise 4.2   Draw a Context Diagram

The context diagram in *Structured Analysis* defines the scope of the system or project. It contains one and only one process. That process represents all sub-processes that are considered part of the system being studied or developed. It also contains external entities that represent other systems, organizations, and entities with which the subject system interfaces. The interfaces are depicted as data flows.
    If you endorse the concept that all the net data flows from the lower-level diagrams (to be drawn later in this lesson) must balance up to the context diagram, then the context diagram can become quite complex. Indeed, it usually becomes so complex that all communication value is lost. Consequently, we choose to restrict the context diagram to show only the key interfaces (most common or important data flows) in the system, deferring less common or critical data flows until we draw the lower-level diagrams. In other words, once introduced, data flows must balance to the lower level diagrams; however, data flows don't necessarily have to balance up to the higher-level and context diagrams. This keeps the upper diagrams readable. Gane and Sarson call data flows that intentionally don't balance upward *trivial data flows*.
    We want you to reproduce the context diagram depicted in Figure 4-2. The main goal of our system is to process subscriptions for membership (which contain special orders), generate monthly promotions for the membership, respond to orders generated from those promotions, and issue order requisitions to the warehouse where they can be filled.

Step 1.   If necessary log on to EXCELERATOR, set the system **HOUSEKEEPING Profile** option, and/or **Restore** your project directory (Lesson 3). Proceed to the Graphics Menu.

Step 2.   Select **Data Flow Diagram.** Select **Add.** Give your graph the name *RECORD CLUB SYSTEM CONTEXT*. Press the Enter key.

Step 3.   From the drawing screen and drawing commands, set your graphics **PROFILE** options the same way you learned in Lesson 3. Do *not* SELECT the **NO ARROW** option since this graph is a true data flow diagram. Use the **PRINT** command to establish your drawing border.

Step 4.   You are now ready to draw your context DFD. **SELECT OBJECT.** Draw the process and external entities. **SELECT CONNECT** to add the data flows. **MOVE** objects and connections to suit your tastes.

Step 5.   Now we are ready to label the processes and data flows. Also, now you are going to begin to see the power of the data dictionary. You may wish to **ZOOM** to **CLOSEUP** to better see the labels. If so, pan to the process in the drawing area.

SELECT **DESCRIBE.** SELECT the process object on your diagram. You are being prompted for an ID. But recall that you have already cataloged this process into the dictionary (when you did your hierarchy chart). There is no need to retype your label (which could result in a typo). You don't even need to remember the ID! Instead of typing in an ID,

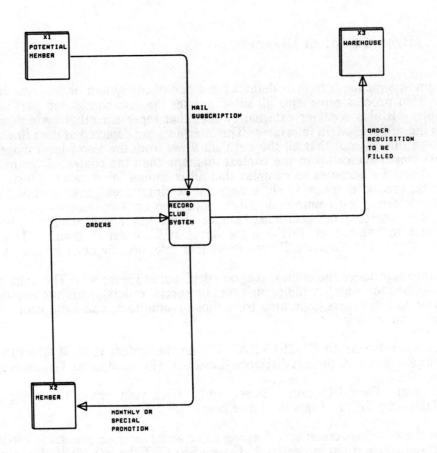

Figure 4-2

Context Data Flow Diagram for the Record Club Project

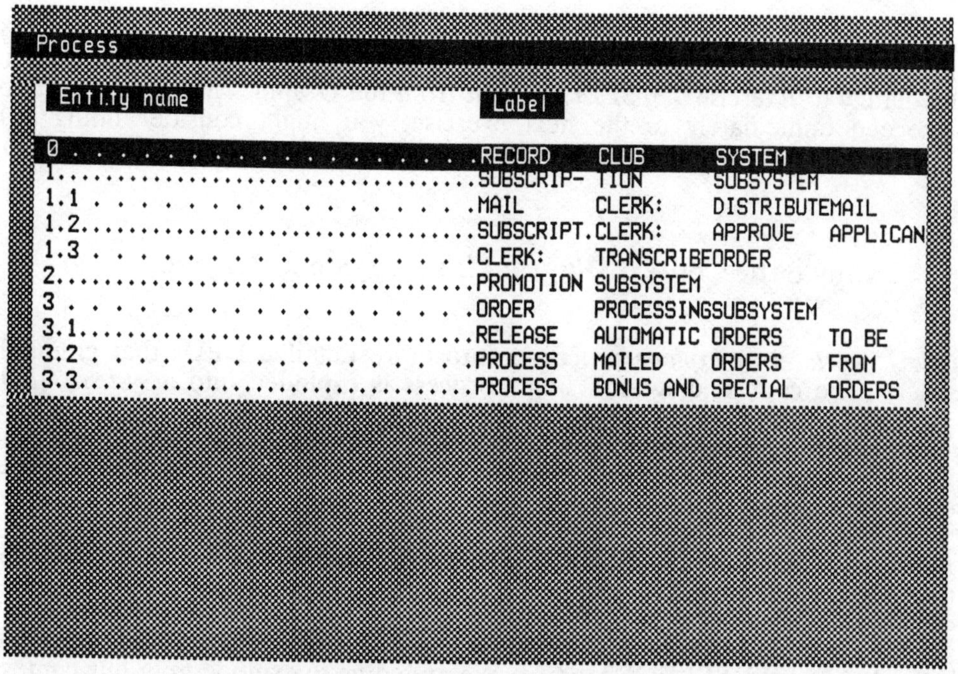

Figure 4-3

Selector List for Existing Data Flow Diagrams

just press the Enter key. You should now see a Selector List (Figure 4-3) containing IDs and labels of all the processes that you DESCRIBEd when you drew the hierarchy chart. This list would *not* have been available to you if you had used the **LABEL** command to name your objects.

Thus, as you can see, there is no need to memorize or create a list of IDs and labels -- so long as you have properly used the **DESCRIBE** command to label objects and take advantage of the dictionary's power. Use your mouse to select Process 0 from that list.

This takes you into the data dictionary entry that you initiated in Exercise 4.1. There is no need to retype the label. However, if you did feel a need to retype the label; the label would automatically change in both the context DFD and in the hierarchy chart. That is another part of the power of the dictionary -- affecting changes throughout your graphs without having to change each graph!

For the time being, just press the F3 function key. Recall that F3 means "save and exit." This takes you back to the graph and the process has been labeled.

Step 6. Next, label the external entities. A few of them were DESCRIBEd to the dictionary back in Lesson 3 when you drew the *SUBSCRIPTION PROCESSING* DFD. Again, use those dictionary entries to generate those data flows on this new DFD. Be sure to give the new external entities abbreviated, unique IDs. We'll definitely want to reuse these external entities on other DFDs; therefore, it is imperative that you use **DESCRIBE** to name these objects.

Step 7. Next, label the data flows. Once again, we'll definitely want to reuse these data flows on other DFDs (to preserve balancing consistency between levels). Therefore, it is imperative that you

49

use **DESCRIBE** to name the data flows.   When prompted   for IDs, make your IDs abbreviations of the labels (we'll explain why in the next exercise - it will prove invaluable!).

Step 8.   That completes Exercise 4.2. **SELECT OTHER. SELECT SAVE. SELECT EXIT** from the drawing commands. **PRINT** the graph if you so desire. If you don't plan to proceed immediately to Exercise 4.3, SELECT **Exit** from the Graphics Menu. Also, if you don't plan to proceed immediately to the next exercise, you might consider doing a **Backup** (via **HOUSEKEEPING**). Printing and backup were covered in Lesson 3, Exercise 3.3.

Exercise 4.3   Explode the Context Diagram into a System Diagram

In *Structured Analysis*, we *level* or *explode* processes into more detailed DFDs that gradually reveal the system to the reader.   The context diagram's single process is exploded into a *system DFD*.   From the hierarchy chart, we already know that we want to explode the context DFD process (number 0) into a system DFD containing three processes (numbers 1, 2, and 3). We should also know that the net data flows to/from Process 0 on the context diagram must balance down to the system diagram. The system diagram we want you to reproduce is depicted in Figure 4-4.

Step 1.   If necessary, log on to EXCELERATOR and/or **Restore** your project directory (Lesson 3). Proceed to the Graphics Menu.

Step 2.   SELECT **Data Flow Diagram.** Even though we want to draw a new data flow diagram, *do not* select **Add.** Instead, SELECT **Modify.** We are going to explode *from* our context DFD. You don't have to remember the name of the context diagram. When prompted for a name, just press the Enter key. A list of your data flow diagrams (including your hierarchy chart) has appeared. Use the mouse to SELECT the context diagram from the list. Your context diagram will reappear in the drawing area. There is no need to reset the **PROFILE** options since they were saved along with the graph.

Step 3.   Before you can explode Process 0 into the system diagram, you must tell EXCELERATOR that you want to. You do this via the data dictionary. SELECT **DESCRIBE.** SELECT the process from the drawing area. The ID should appear in the Status Line. Since we don't want to change the ID, press the Enter key. This takes you back to the data dictionary entry for the process (Figure 4-5).

Notice that there is an **Explodes To One Of** area for describing *what* the process explodes to. An entry made in this area is said to define the *explosion path* for the object -- in this case a process. We want to explode the process to a system DFD; therefore SELECT the first blank position in that block.   Here, we must enter the name of the DFD to which we will explode. Type *RECORD CLUB SYSTEM DIAGRAM*. Press the F3 function key to return to the context DFD.

Step 4.   Now we can explode the context process to reveal greater detail.   **SELECT EXPLODE.** SELECT the process from the graph. A blank drawing screen will appear. The name and message *LEVEL 2* will appear above the Orientation Map. A link between the two graphs has been established in your dictionary.

Step 5.   Set your **PROFILE** options the same way you learned in Lesson 3. Use the **PRINT** command to establish your drawing border.

Step 6.   You are now ready to draw the system DFD. SELECT **OBJECT.** Draw the processes and external entities. SELECT **CONNECT** to add the data flows. **MOVE** objects and connections to suit your tastes.

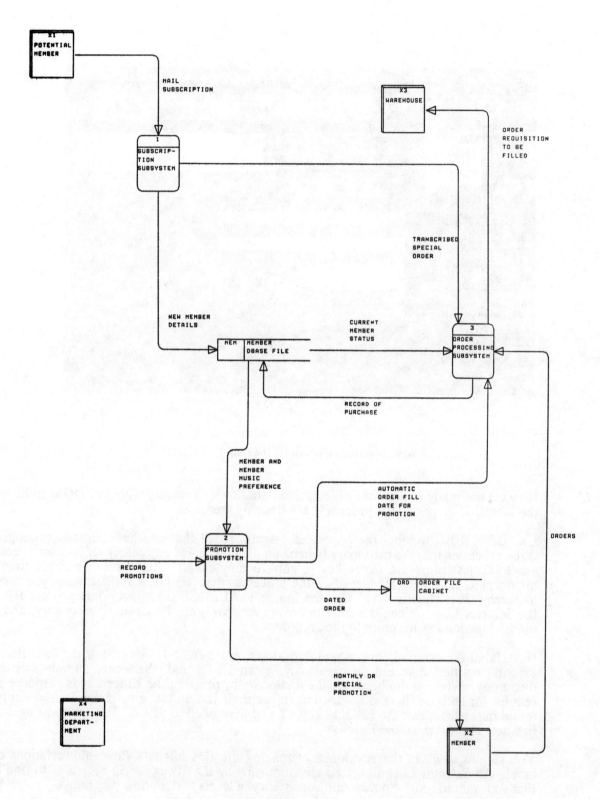

Figure 4-4

System Data Flow Diagram for Record Club Project

51

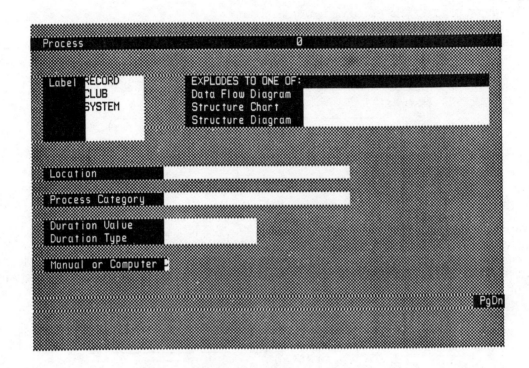

Figure 4-5

Process Description Screen (in the Data Dictionary)

Step 7.   Now we are ready to label the objects and data flows. You may wish to **ZOOM** to better see the labels. If so, pan to the process in the drawing area.

Use **DESCRIBE** to label the processes. Again, recall that you have already cataloged the three processes into the dictionary (when you did your hierarchy chart). If you don't recall the exact ID, just press the Enter key (highly recommended - much more reliable than your memory) to get a Selector List of all the processes that you DESCRIBEd when you drew the hierarchy chart (Figure 4-1). Use the mouse to SELECT the appropriate process IDs from the Selector Lists. When the data dictionary appears, edit the **Label** (if necessary) and press the F3 function key to return to the graph.

Step 8.   Next, label the external entities and data flows on Figure 4-4. Several of the data flows and external entities appeared on previously drawn DFDs and, therefore, already exist in the dictionary. When in doubt, check the dictionary by pressing the Enter key in response to the request for an ID. There is no need to retype their IDs and labels. If the desired entity isn't found in the list, press the ESCape key. You'd then need to SELECT the process again, type in a new ID, and press the Enter key.

You should now see the advantage of making the IDs for data flows abbreviations of the label. All Selector Lists all sorted alphabetically by ID. Thus, it will be easier to find labels that you want to reuse (to maintain consistency in levels of data flow diagrams).

For external entities and data flows not already in your dictionary, be sure to use the **DESCRIBE** command to name them.

52

Step 9. Next, name the data stores. Once again, you created two of the data stores back in Lesson 3. Both data stores are still in the dictionary and should be named from the existing data dictionary entry. Give the rest of the data stores IDs and labels as shown in Figure 4-4. Once again, we'll definitely want to reuse these names on other DFDs. Therefore, it is imperative that you use **DESCRIBE** to name the data stores.

Step 10. That completes Exercise 4.3. **PRINT** the graph if you so desire. SELECT **EXIT** from the drawing commands. Now SELECT **SAVE** (EXCELERATOR will save the *system* and *context* diagrams.) SELECT **Exit** from the Graphics Menu. Also, if you don't plan to proceed immediately, you might consider doing a **Backup** (via **HOUSEKEEPING**). Printing and backup were covered in Lesson 3.

Exercise 4.4   Explode a Process into a Previously Drawn DFD

Structured Analysis proposes a top-down strategy for developing a leveled set of data flow diagrams. However, in practice, you might choose to compose initial drafts of some lower-level diagrams prior to drawing the upper levels. Why? Although *Structured Analysis* is a top-down technique, facts generally reveal themselves from the bottom up (details first). EXCELERATOR supports both top-down and bottom-up techniques. In fact, you've already developed one bottom-level diagram for the Record Club system -- back in Lesson 3 (*SUBSCRIPTION PROCESSING*). Let's link that bottom-level diagram into our top-down set of DFDs. All you have to do is link the DFD to the *system diagram*.

Step 1. If necessary, log on to EXCELERATOR, set your **HOUSEKEEPING Profile**, and/or **Restore** your project directory (Lesson 3). Proceed to the Graphics Menu.

Step 2. Do you remember the name of the DFD you drew in Lesson 3? Before you turn back to Lesson 3, you don't have to remember the name! SELECT **List** from the Action Keypad. When prompted for a *NAME*, just press the Enter key. A Selector List of DFDs is displayed. For later reference, *write* down the name of the *SUBSCRIPTION PROCESSING DFD* that you drew in Lesson 3.

Step 3. *When drawing a top-down, leveled set of DFDs, we recommend that you always enter the set through its context diagram!* Why? Because EXCELERATOR establishes the top level based on the point at which you entered the graphs. By selecting the context diagram as this point, you get total freedom to move from level to level via any graph in the leveled set.

SELECT **Modify** from the Action Keypad for **Data Flow Diagram**. Press the Enter key and then SELECT the context diagram from the list of DFDs. The context diagram should reappear on your screen. There is no need to reset the drawing command **PROFILE** options for an existing graph.

Step 4. SELECT **EXPLODE** from the drawing commands. SELECT the sole process on the context diagram. The context diagram should be replaced by the diagram you drew in the last exercise.

Step 5. Now you are ready to explode (actually link) Process 1 on the system diagram to the DFD you drew in Lesson 3. You do this via the data dictionary. SELECT **DESCRIBE**. SELECT Process 1 from the drawing area. The ID should appear in the Status Line. Since we don't want to change the ID, press the Enter key. This takes you back to the data dictionary entry for Process 1. We want to explode the process to the *SUBSCRIPTION PROCESSING* DFD you created in Lesson 3. Therefore, move the cursor from the **Label** area to the **Explodes To Data Flow Diagram** line (again, you can use either the mouse or the Tab key to move the cursor). Type *SUBSCRIPTION PROCESSING*, the name you wrote down when you listed the DFDs in Step 1. Press the F3 function key to return to the system diagram.

Step 6. Let's make sure it worked! SELECT **EXPLODE**. SELECT Process 1 from the drawing area. If the link was successful, the *SUBSCRIPTION PROCESSING* DFD that you drew in Lesson 3 should appear in the drawing area. The name and message *LEVEL 3* will appear above the Orientation Map. The message *LEVEL 3* indicates that you are currently viewing a data flow diagram that is linked to other DFDs through the data dictionary.

Step 7. The advantage of linked DFDs can now be demonstrated. The different levels represent different levels of abstraction, general to detailed. You are currently looking at a detailed DFD. Suppose you want to see how subscription processing fits into the overall system. There is no need to exit this diagram. SELECT **OTHER**. From the sub-menu, SELECT **RETURN**. This command takes you one level higher, allowing you to review the overall system. If you were to execute the command again, it would take you to the context diagram. Instead, SELECT **EXPLODE**. Now, SELECT Process 1 again. This will return you to Level 3, the subscription processing subsystem.

Step 8. EXCELERATOR will allow you to explode up to ten levels deep! Suppose you are looking at a DFD at Level n where n is greater than 2. EXCELERATOR allows you to quickly skip the intermediate levels on your way to level 1. **Let's try.** SELECT **OTHER**. From the sub-menu, SELECT **RETRNTOP**. This will take you back to the context diagram. As you can see, moving between levels of abstraction is very easy.

One of the problems with DFDs is maintenance. It is very frustrating to spend hours drawing DFDs with paper and pencil only to learn, after reviewing them with your users, that they are incomplete or incorrect. Once again, EXCELERATOR greatly simplifies this task. We just linked up a subscription processing DFD.

Let's now assume the inevitable -- the users forgot to tell us all the important details. Modify the subscription processing DFD according to Figure 4-6. You should be able to make most of the modifications with the exception of the diverging data flow coming from Process 1.4 and the data flow, from Process 1.3, which appears to be going nowhere. These new techniques are described as follows:

Step 9. Diverging data flows are really quite simple. Just draw the data flows separately. Then **MOVE** one data flow's exit port to precisely match the other data flow's exit side and port location (or move both data flows to to a common, new side and port location). You learned how to move data flows in Lesson 3.

Step 10. The data flow *TRANSCRIBED SPECIAL ORDER*, from Process 1.3, is a balancing data flow. We are assuming that you are familiar with the concept of balancing data flow diagrams. Balancing ensures consistency with the previous, higher level. Reexamine the system diagram, Figure 4-4. Notice that the data flow *TRANSCRIBED SPECIAL ORDER* exits Process 1, the subscription subsystem, and enters Process 3, the order subsystem. Now that we've exploded Process 1 to greater detail (Figure 4-6), we can't lose that data flow. But we don't want to include details about Process 3 on the detailed DFD for Process 1. Thus, we introduce a balancing data flow as follows:

    a. First we need to **DELETE** the existing data flow labeled *TRANSCRIBED SPECIAL ORDER* and the connected entity *SPECIAL ORDERS GROUP*. Recall that we've determined that the data flow should actually go to a process, rather than an entity. SELECT **DELETE**. SELECT the *SPECIAL ORDERS GROUP* entity. The Status Line now says *All connections to this object will be deleted. Proceed? (Y/N)*. Since we also want the data flow to be DELETEd type Y. The status line now asks if you want the dictionary description of the entity to be DELETEd. **CAREFUL!** Since the entity may currently exist elsewhere some other DFD or may be reused later, SELECT **NO** from the submenus. Now you're asked whether the data flow should be DELETEd from the dictionary. We'll certainly want to reuse it so SELECT **NO**. The entity and data flow should now be deleted.

    b. Now let's draw the *TRANSCRIBED SPECIAL ORDER* data flow as depicted in Figure 4-6. SELECT **INTRFACE** from the drawing commands. A sub-menu appears below the graphics commands. Since we want to create a net output data flow, SELECT

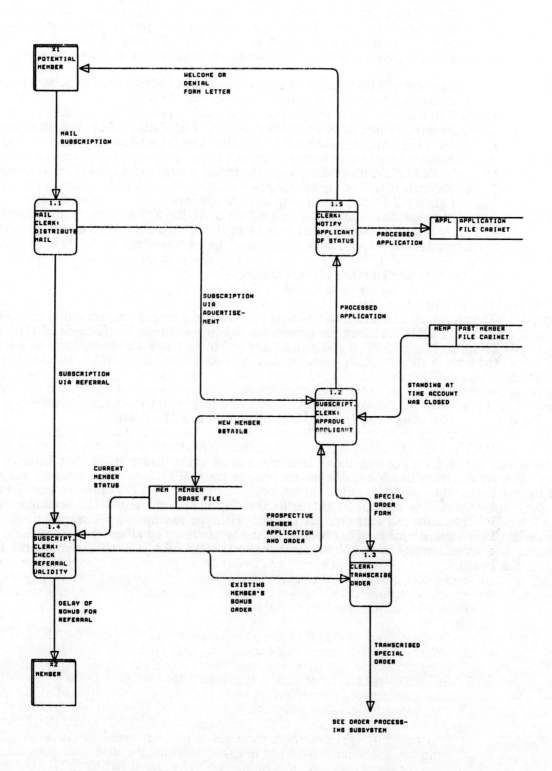

Figure 4-6

Modified Subscription Processing Data Flow Diagram

OUTPUT. The Status Line is instructing you select a *Location to start interface*. Respond by SELECTing the location for the end of the data flow. A handle appears at that location. The Status Line now instructs you to *Touch a second obj or an intermediate location*. Respond by SELECTing Process 1.3. Now SELECT the desired side and port location of the process. The balancing data flow now appears.

c.   Now **DESCRIBE** the Interface. An Interface **DESCRIBE**s as a data flow in the data dictionary.

d.   In pure *Structured Analysis*, you'd leave it at that. But a data flow going nowhere quickly loses it's communication value to users who would have to wade through a number of DFDs to find the destination of the data flow. Therefore, we suggest that you add a comment that directs the reader where to look for the destination. SELECT **TXT BLOCK**. Define an imaginary rectangle by SELECTing the upper-left and lower-right hand corners. A pop-up window appears in the center of the screen. Type in the message *SEE ORDER PROCESSING SUBSYSTEM*. Press the Enter key. The message has been added next to (but not linked to) the balancing data flow. The reader can respond to the message in one of two ways:

o   Look for the system diagram
o   Look for the detailed DFD for order processing

Step 11.   That completes Exercise 4.4. Unless you want to proceed immediately to the next exercise, SELECT **EXIT** and then **Save** from the drawing commands. Then SELECT **Exit** from the Graphics Menu. Also, if you plan to quit and leave the workstation, consider backing up your data.

Exercise 4.5   Complete the Leveled Set of Data Flow Diagrams

You know what you need to know to draw a leveled set of data flow diagrams. In Part Three of this tutorial, you will discover that EXCELERATOR can also analyze your DFDs for consistency and completeness. In order to prepare for those lessons, and to further practice exploding DFDs, draw the remaining graphs for the record club system. The new DFDs are depicted in Figures 4-7 through 4-11. Be sure to practice what you've learned. Especially, use your dictionary -- NEVER type any label for a second time. You'll eventually realize the ultimate benefit. That benefit is that by not retyping labels, your data flows will balance from level to level, generating fewer (or no) errors when EXCELERATOR checks your work for errors (in Part II of this book).

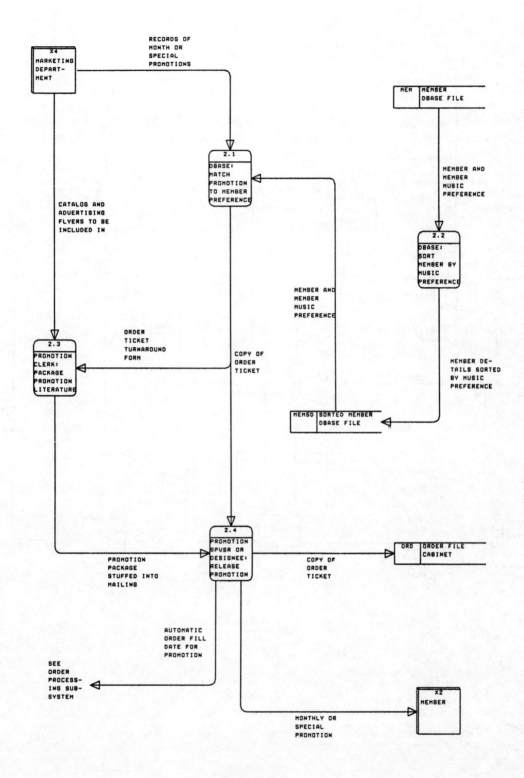

Figure 4-7

Data Flow Diagram

57

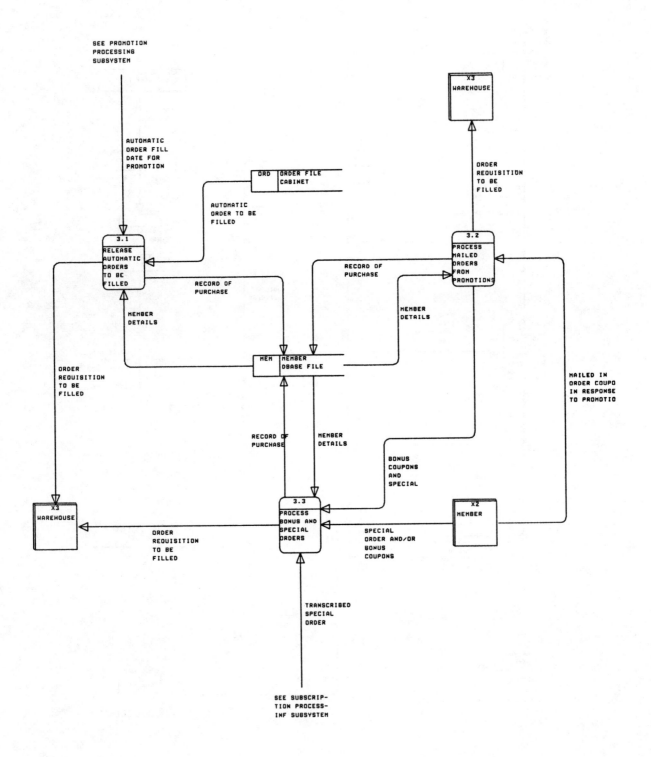

SEE PROMOTION
PROCESSING
SUBSYSTEM

X3
WAREHOUSE

AUTOMATIC
ORDER FILL
DATE FOR
PROMOTION

ORDER
REQUISITION
TO BE
FILLED

ORD | ORDER FILE
CABINET

AUTOMATIC
ORDER TO BE
FILLED

3.1
RELEASE
AUTOMATIC
ORDERS
TO BE
FILLED

3.2
PROCESS
MAILED
ORDERS
FROM
PROMOTIONS

RECORD OF
PURCHASE

RECORD OF
PURCHASE

MEMBER
DETAILS

MEMBER
DETAILS

MEM | MEMBER
DBASE FILE

ORDER
REQUISITION
TO BE
FILLED

MAILED IN
ORDER COUPO
IN RESPONSE
TO PROMOTIO

RECORD OF
PURCHASE

MEMBER
DETAILS

BONUS
COUPONS
AND
SPECIAL

X3
WAREHOUSE

3.3
PROCESS
BONUS AND
SPECIAL
ORDERS

X2
MEMBER

ORDER
REQUISITION
TO BE
FILLED

SPECIAL
ORDER AND/OR
BONUS
COUPONS

TRANSCRIBED
SPECIAL
ORDER

SEE SUBSCRIP-
TION PROCESS-
INF SUBSYSTEM

Figure 4-8

Data Flow Diagram

58

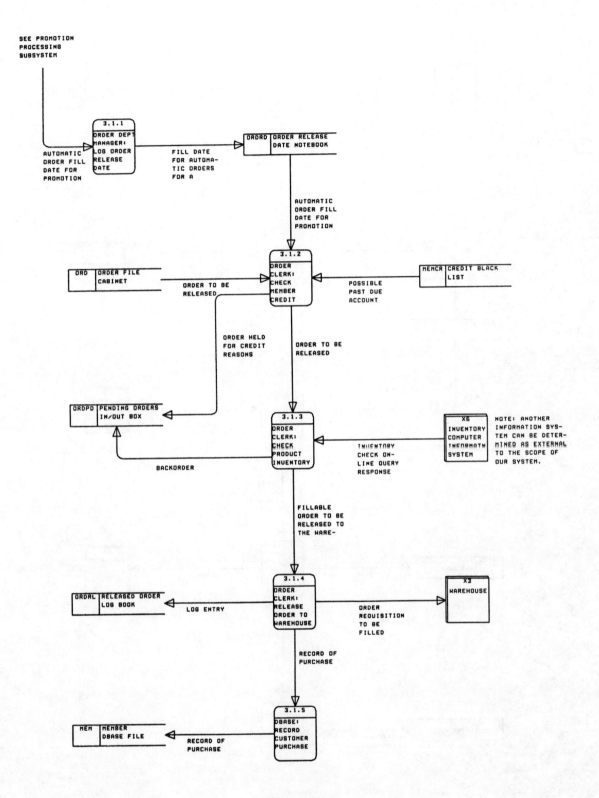

Figure 4-9

Data Flow Diagram

59

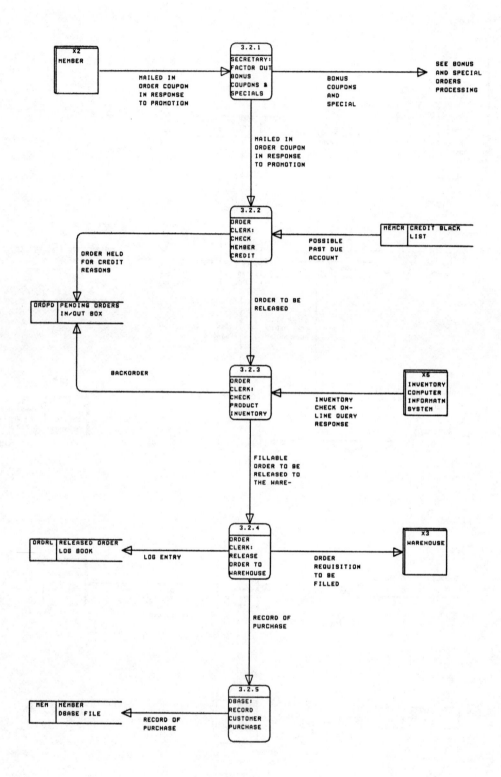

Figure 4-10

Data Flow Diagram

60

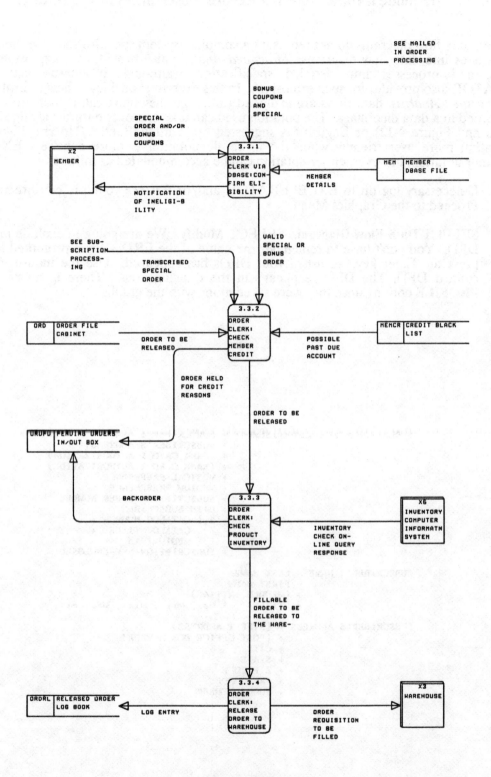

Figure 4-11

Data Flow Diagram

61

Exercise 4.6   Explode a Data Flow to a Record Description in the Data Dictionary

By themselves, data flow diagrams do not represent a complete system specification. You've learned how to explode processes into data flow diagrams. *Structured Analysis* also provides for explosion of data flows, data stores, and processes into detailed specifications (although it doesn't call it explosion). EXCELERATOR also provides for such explosions. In this exercise, you'll learn how to explode data flows.

     In *Structured Analysis*, data flows are exploded (although they don't call it that) into data structures that are recorded in a data dictionary. The notation used can take one of two forms: algebraic (as suggested by DeMarco and Figure 4-12) or English (as suggested by Whitten/Bentley/Ho and Figure 4-13). The English notation more user friendly while the algebraic notation is more concise. EXCELERATOR supports neither notation; however, either notation can be accommodated as follows:

Step 1.   If necessary log on to EXCELERATOR and/or **Restore** your project directory (Lesson 3). Proceed to the Graphics Menu.

Step 2.   SELECT **Data Flow Diagram**. SELECT **Modify**. We are going to explode *from* our context DFD. You don't have to remember the name of the ERD. When prompted for a name, just press the Enter key. A list of your DFDs has appeared. Use the mouse to SELECT the context DFD. The DFD reappear's in the drawing area. There is no need to reset the **PROFILE** options since they were saved along with the graph.

```
SUBCRIPTION VIA ADVERTISEMENT = SUBSCRIBER'S NAME:        * key
                              + SUBSCRIBER'S ADDRESS:
                              + (BANK CARD 1 AUTHORIZATION)
                              + (BANK CARD 2 AUTHORIZATION)
                              + MUSICAL PREFERENCE
                              + MEDIUM PREFERENCE
                              + SUBSCRIPTION OFFER NUMBER
                              + DATE SUBSCRIBED
                              + 3 { CATALOG NUMBER +
                              +     CATALOG TITLE   +
                                    MEDIUM} 15
                              + SUBSCRIPTION FEE ENCLOSED

SUBSCRIBER'S NAME = LAST NAME
                  + FIRST NAME
                  + (MIDDLE INITIAL)
                  + (TITLE)    * Mr., Mrs., Miss, Ms., etc.

SUBSCRIBER'S ADDRESS = (STREET ADDRESS)
                     + (POST OFFICE BOX NUMBER)
                     + CITY
                     + STATE
                     + ZIP CODE
                     + AREA CODE
                     + PHONE NUMBER
```

Figure 4-12

Algebraic Data Structure Notation (DeMarco)

```
An occurrence of SUBSCRIPTION VIA ADVERTISEMENT consists of the
following:
        SUBSCRIBER'S NAME which consists of the following:
             LAST NAME
             FIRST NAME
             MIDDLE INITIAL (optional)
             TITLE (optional)
        SUBSCRIBER'S ADDRESS which consists of the following:
             STREET ADDRESS (optional)
             POST OFFICE BOX NUMBER (optional)
             CITY
             STATE
             ZIP CODE
             AREA CODE
             PHONE NUMBER
        BANK CARD 1 AUTHORIZATION (optional)
        BANK CARD 2 AUTHORIZATION (optional)
        MUSICAL PREFERENCE
        MEDIUM PREFERENCE
        SUBSCRIPTION OFFER NUMBER
        DATE SUBSCRIBED
        3 to 15 occurrences of SUBSCRIBER'S ORDER ITEM GROUP which
        consists of the following:
             CATALOG NUMBER
             CATALOG TITLE
             MEDIUM
        SUBSCRIPTION FEE ENCLOSED
```

Figure 4-13

English Data Structure Notation (Whitten/Bentley/Ho)

Use the **EXPLODE** command to get to the *SUBSCRIPTION PROCESSING* DFD.

Step 3. As was the case with processes, we can't explode data flows until their explosion path has been **DESCRIBE**d in the dictionary. SELECT **DESCRIBE** from the graphics commands. SELECT the data flow *SUBSCRIPTION VIA ADVERTISEMENT* from the DFD. The Status Line is displaying the ID. Since we don't want to change the ID, press the Enter key. This takes you back to the data dictionary entry for the data flow (Figure 4-14).

Notice that there is an **Explodes To One Of** area for describing *what* the data flow explodes to. We want to explode the data flow to a data **Record**; therefore, SELECT that line, the first blank position. Here, we must enter the name of the record to which we will explode. For data flows, we prefer to give them the identical name of the data flow from which they are exploded. For instance, we will explode the *SUBSCRIPTION VIA ADVERTISEMENT* data flow into the *SUBSCRIPTION VIA ADVERTISEMENT* record layout. Don't worry! EXCELERATOR won't get confused. It identifies objects by a combination of object type (e.g., record) and name (e.g., *SUBSCRIPTION VIA ADVERTISEMENT*). The other

63

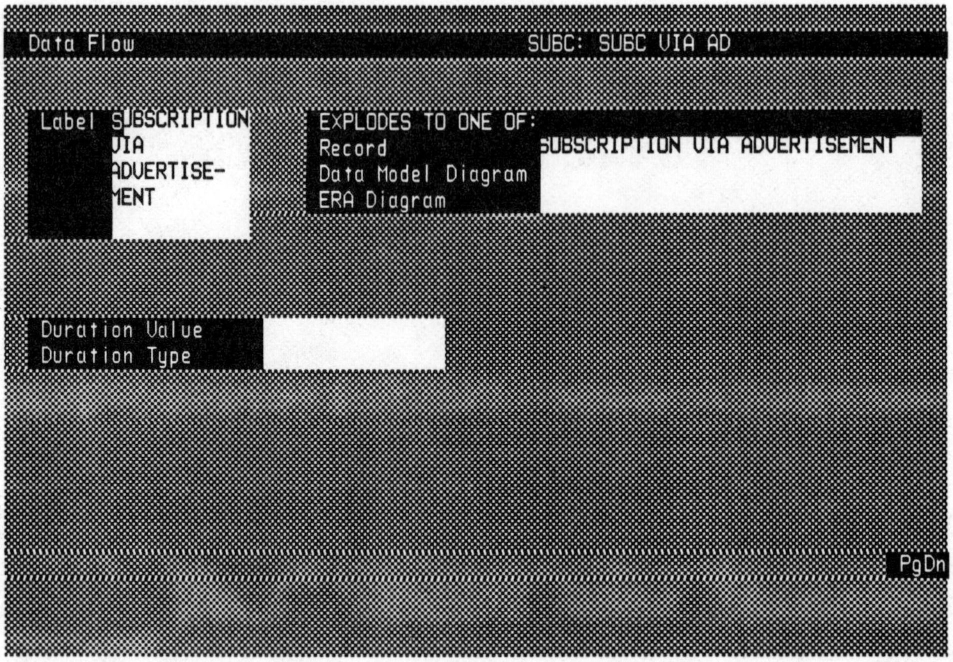

Figure 4-14

Data Flow Description Screen (in the Data Dictionary)

occurrence of *SUBSCRIPTION VIA ADVERTISEMENT* has a different type, data flow. Type the name of the record (entity). Press the F3 function key to save and exit the dictionary. You have returned to the graph.

Step 4.   Now we can explode the data flow. **SELECT EXPLODE.** SELECT the *SUBSCRIPTION VIA ADVERTISEMENT* data flow from the graph. A blank Record Description Screen (Figure 4-15) will appear. The sections of this screen are described as follows:

o   **Alternate Name.** This is used to record an alias.
o   **Definition.** This is a one-line description field.
o   **Name of Element or Record.** This is used to record the names of elements or groups of elements (also known as a sub-record). This is a scroll region that can accommodate up to 115 entries.
o   **Occ.** This stands for *number of occurrences of the element or sub-record in a single occurrence of the record.*
o   **Seq.** This stands for *sequence.* The numbers you place here determine the sequence of the elements and sub-records in the record. Thus, you can insert new elements or sub-records at the end of the record and alter the sequence such that they will print in the desired sequence.
o   **Type.** This field is used to associate a type with the corresponding data element or sub-record. We have adopted the following internal legend for data flows:

    E   Data element
    R   Sub-record: group of elements that always occur together
    9   An optional, non-key data element

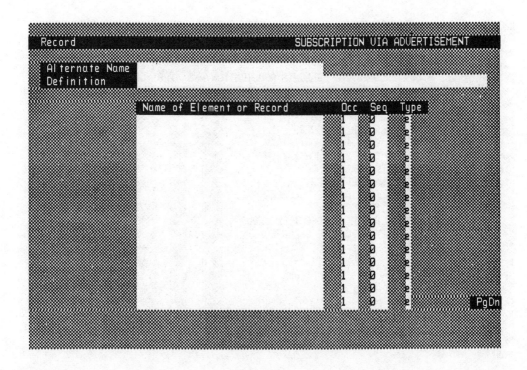

Figure 4-15

Data Record Description Screen (in the Data Dictionary)

Step 5. Now we can begin to define the data structure for the data flows. Figure 4-16 illustrates the EXCELERATOR notation for the *SUBSCRIPTION VIA ADVERTISEMENT* data structure that was depicted in Figures 4-12 and 4-13. Use the mouse to SELECT the first line in the **Name of Element or Record** block. Begin entering the elements and sub-records. Notice how we handled groups and elements. After you type the **Name of Element or Record** field, press the Enter key to move to the **Occ.** field. The default value is *1*. Change it or press Enter to move to the **Seq.** field. Change the default value or press Enter to move to the **Type** field. The default is *E*. Change it or press Enter, this time moving to the next line in the **Name of Element or Record** field. You can use the up-arrow and down-arrow keys to move up and down in the scroll region.

Step 6. You may be wondering how to define the data elements that make up the sub-records (**Type** = *R*)? It's very easy. Use the mouse to position the cursor somewhere in the *SUBSCRIBER'S ORDER ITEM GROUP* sub-record. We want to **DESCRIBE** that record in the data dictionary. Unlike the graphics facility, there is no **DESCRIBE** command to SELECT from the screen. But there is a **DESCRIBE** key, the F4 function key. Press the F4 function key.

This takes you to a new Record Description screen, specifically for the *SUBSCRIBER'S ORDER ITEM GROUP*. Complete the layout per Figure 4-17. After completing the sub-record, press the F3 save and exit function key to return to the parent record. Press the F3 key again to save and exit to the data flow diagram. Now **DESCRIBE** the *SUBSCRIBER'S NAME* and *SUBSCRIBER'S ADDRESS* records.

Step 7. Some additional record layouts for data flows are illustrated in Figure 4-18, some in algebraic form and others in English form. Convert them to EXCELERATOR descriptions that are linked to the DFD. When you are finished, return to the DFD.

65

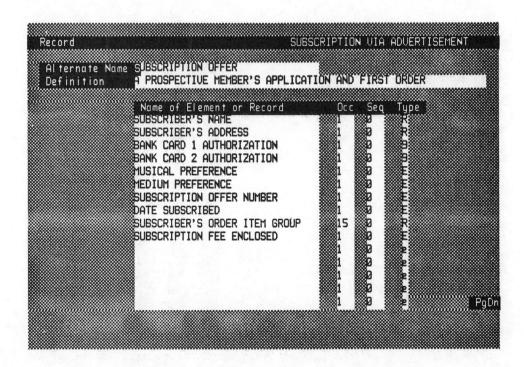

Figure 4-16

Data Record Description Screen Illustrating Data Structure

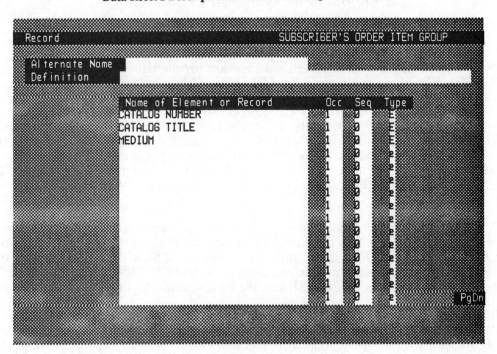

Figure 4-17

Data Record Description Screen for a Sub-Record or Group of Data Elements

An occurrence of PROCESSED APPLICATION consists of the following:
    MEMBER NUMBER (optional)
    SUBSCRIBER'S NAME which consists of the following:
        LAST NAME
        FIRST NAME
        MIDDLE INITIAL (optional)
        TITLE (optional)
    SUBSCRIBER'S ADDRESS which consists of the following:
        STREET ADDRESS (optional)
        POST OFFICE BOX NUMBER (optional)
        CITY
        STATE
        ZIP CODE
        AREA CODE
        PHONE NUMBER
    BANK CARD 1 AUTHORIZATION (optional)
    BANK CARD 2 AUTHORIZATION (optional)
    MUSICAL PREFERENCE
    MEDIUM PREFERENCE
    SUBSCRIPTION OFFER NUMBER
    DATE SUBSCRIBED
    3 to 15 occurrences of SUBSCRIBER'S ORDER ITEM GROUP which
    consists of the following:
        CATALOG NUMBER
        CATALOG TITLE
        MEDIUM
    SUBSCRIPTION FEE ENCLOSED
    APPLICATION STATUS

An occurrence of STANDING AT TIME ACCOUNT WAS CLOSED consists of the
following:
    MEMBER NAME
    MEMBER NUMBER
    MEMBER ADDRESS
    DATE OPENED
    DATE CLOSED
    ACCOUNT STATUS
    ACCOUNT BALANCE

SPECIAL ORDER FORM = MEMBER NUMBER   * key
                  + 3 { CATALOG NUMBER +
                        CATALOG TITLE +
                        MEDIUM } 15

Figure 4-18(a)

Data Structures for Other Data Flows (Part 1)

67

```
PROSPECTIVE MEMBER APPLICATION AND ORDER = SUBSCRIBER'S NAME:      * key
                                         + SUBSCRIBER'S ADDRESS:
                                         + (BANK CARD 1 AUTHORIZATION)
                                         + (BANK CARD 2 AUTHORIZATION)
                                         + MUSICAL PREFERENCE
                                         + MEDIUM PREFERENCE
                                         + SUSCRIPTION OFFER NUMBER
                                         + DATE SUBSCRIBED
                                         + 3 { CATALOG NUMBER +
                                               CATALOG TITLE   +
                                               MEDIUM } 15
                                         + SUBSCRIPTION FEE ENCLOSED

MEMBER DETAILS = MEMBER NUMBER      * key
               + MEMBER NAME
               + DATE ENROLLED
               + (BANK CARD 1 AUTHORIZATION)
               + (BANK CARD 2 AUTHORIZATION)
               + MEMBER ADDRESS:
               + MUSICAL PREFERENCE
               + MEDIUM PREFERENCE
               + MEMBER BALANCE DUE
               + ACCOUNT STATUS
               + MEMBERSHIP PURCHASE REQUIREMENT
               + MEMBERSHIP EXPIRATION DATE
               + MEMBER PURCHASES TOWARD REQUIREMENT
               + MEMBER PURCHASES BEYOND REQUIREMENT
               + BONUS CREDITS CLAIMED
```

Figure 4-18(b)

Data Structures for Other Data Flows (Part 2)

Step 8.   SELECT **OTHER**. SELECT **SAVE**. SELECT **EXIT**. This should take you from your ERD graph to the Graphics Menu. SELECT **Exit** to leave the **GRAPHICS** facility. Now you are at the Main Menu.

Step 9.   You cannot print your data dictionary descriptions from the **GRAPHICS** facility. If you want a hard copy, do the following (otherwise, skip to Step 10). SELECT **XLDICTIONARY**. This takes you to an entirely new facility and screen (Figure 4-19). From this **XLDICTIONARY** menu, SELECT **REC/ELE** (which stands for *RECords and ELEments*). A sub-menu appears in the middle of the screen. SELECT **Record**. An Action Keypad appears on the right side of the screen (Figure 4-20). For this exercise, ignore that large sub-menu in the middle of the screen.

SELECT **Output** from the Action Keypad. The **Name** prompt will appear. We want *all* records; therefore, type an asterisk (the wildcard character) and press the Enter key. The

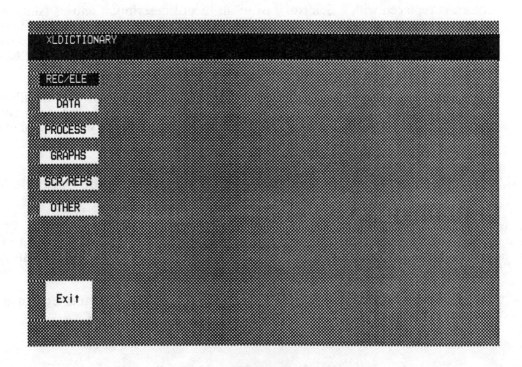

Figure 4-19

Initial XLDICTIONARY Menu Screen

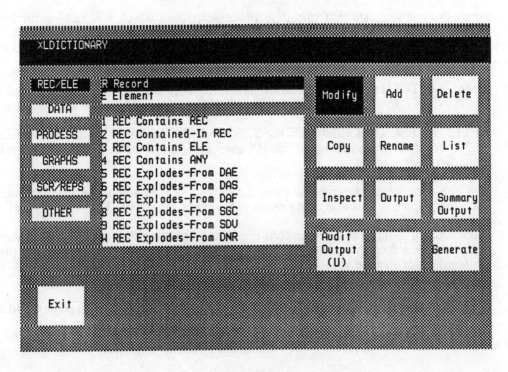

Figure 4-20

Action Keypad for XLDICTIONARY Menu Screen

screen is replaced with a Selector List of all record description entries (similar to Figure 4-21). SELECT the first item on the list, labeled *All Entities on Selector List*. Below the Action Keypad you will see your output options. SELECT **Printer** (this step assumes that your PC is connected to a supported printer). After your output prints, SELECT **Exit** to return to the Main Menu.

Step 10.   That completes Exercise 4.7. Data elements can be defined in greater detail; however, that technique is more fully described in Lesson 5. If you don't plan to proceed immediately, you might consider doing a **Backup** (via **HOUSEKEEPING**). Backup was covered in Lesson 3.

Exercise 4.7   Explode a Data Store to a Record Description in the Data Dictionary

Structured Analysis also provides for explosion of data stores into detailed specifications. If you want to describe data stores to records that describe the store's structure, the technique is identical to the one you learned in the last exercise. Just substitute data store for data flow. Try it! First **DESCRIBE** the *MEMBER DBASE FILE* data store as exploding to a record layout. **EXPLODE** the data store according to the data dictionary entry in Figure 4-22.

     Alternatively, you can explode data stores to data models. We will defer this technique until Lesson 5 which specifically covers data modeling documentation.

Exercise 4.8   Explode a Primitive Process into a Mini-Spec Description in the Data Dictionary

Structured Systems Analysis suggests that primitive processes, those which will not be further exploded to more detailed DFDs, be documented as process narratives or *mini-specs*. Mini-specs use a logic language called *Structured English* to specify the procedure that transforms the process's inputs into its outputs. Mini-specs can easily be added to the EXCELERATOR structured specification as follows:

Step 1.   Enter your DFDs through the context diagram. **EXPLODE** the processes until you get to the *SUBSCRIPTION PROCESSING* DFD. All of the processes on this diagram are primitive.

Step 2.   SELECT **DESCRIBE**. SELECT the *SUBSCRIPT. CLERK: APPROVE APPLICANT* process (1.2). We don't want to change the ID; therefore, press the Enter key. This takes you to the data dictionary screen.

Step 3.   For the time being, ignore the entry fields you see on the screen. Like many other data dictionary screens, this one contains multiple pages. Press the PgDn key to look at page 2.

Step 4.   Now you are looking at a screen that contains a **Description** block. The purpose of the block is to provide a free-format mechanism to record additional details about the process. This is an ideal place to type your mini-spec. Try it! Type the mini-spec illustrated in Figure 4-23. When you're finished, press the F3 function key to save and exit.

Step 5.   This completes Lesson 4. Before you exit EXCELERATOR, we recommend that you back up your work.

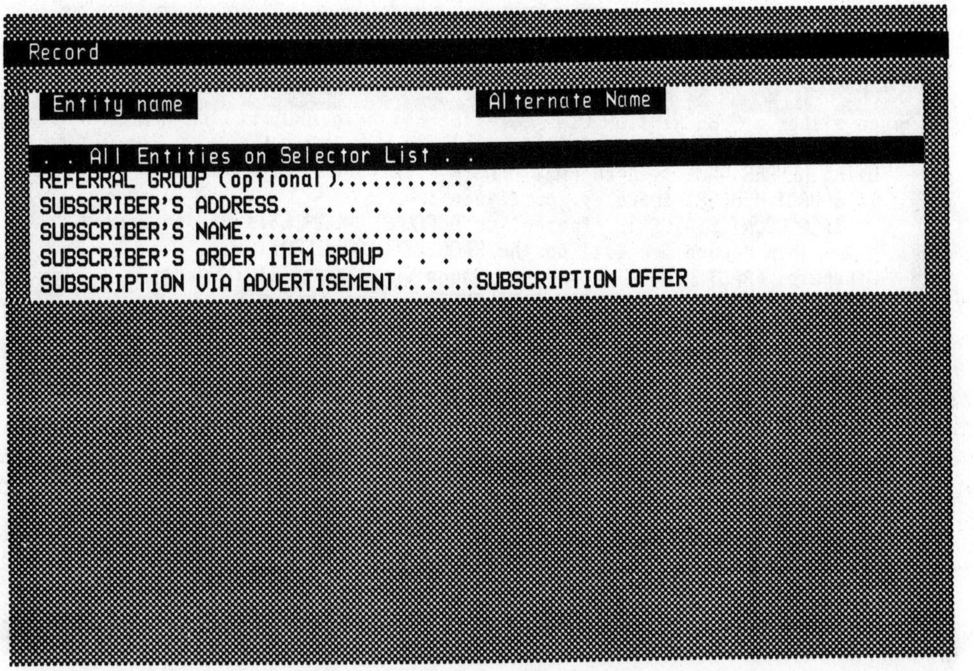

Figure 4-21

Selector List for Records Described in the Dictionary

```
A MEMBER record consists of the following:
      MEMBER NUMBER
      MEMBER NAME
      DATE ENROLLED
      BANK CARD 1 AUTHORIZATION (optional)
      BANK CARD 2 AUTHORIZATION (optional)
      MEMBER ADDRESS which consists of the following:
          STREET ADDRESS (optional)
          POST OFFICE BOX NUMBER (optional)
          CITY
          STATE
          ZIP CODE
          AREA CODE
          PHONE NUMBER
      MUSICAL PREFERENCE
      MEDIUM PREFERENCE
      MEMBER BALANCE DUE
      ACCOUNT STATUS
      MEMBERSHIP PURCHASE REQUIREMENT
      MEMBERSHIP EXPIRATION DATE
      MEMBER PURCHASES TOWARD REQUIREMENT
      MEMBER PURCHASES BEYOND REQUIREMENT
      BONUS CREDITS CLAIMED
```

Figure 4-22

Data Structure for a Data Store's Record

71

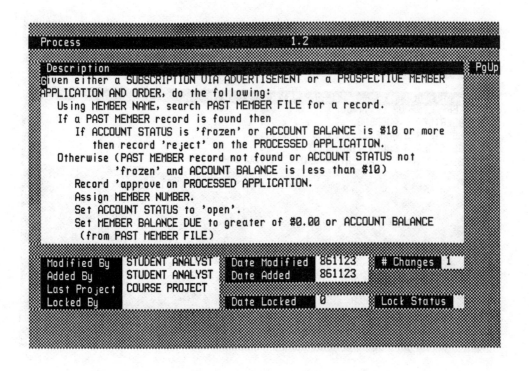

Figure 4-23

Mini-Spec Recorded in a Process Description Screen

Conclusion

In this lesson you learned to use EXCELERATOR to support *Structured Systems Analysis* documentation. Using the **GRAPHICS** facility, you draw a leveled set of data flow diagrams that model the system, beginning with the current system and proceeding to the target system. For detailed specifications, data flows and data stores can be exploded into record descriptions (or data models, which are taught in the next lesson). To complete the detailed analysis specifications, the primitive processes on DFDs are described as mini-specs in the **Description** block of the process data dictionary forms.

As a side note, you may be aware that *Structured Systems Analysis* has a companion methodology, *Structured Systems Design*. The companion methodology transforms data flow diagrams from *Structured Analysis* into hierarchical structure charts that represent an ideal modular design for the computer programs to be written. In EXCELERATOR, you can explode any DFD process into either a Yourdon-style or Jackson-style structure chart, thereby completing the structured systems specification. Individual modules on the structure charts can be exploded into module descriptions or logic that can be passed along to the programmers. Thus, EXCELERATOR supports the entire structured methodology.

Structured Analysis is a data flow approach. Many organizations, especially those who do numerous database applications, have complemented *Structured Analysis* with a data modeling methodology. One such methodology, *Information Engineering*, will be introduced in the next lesson.

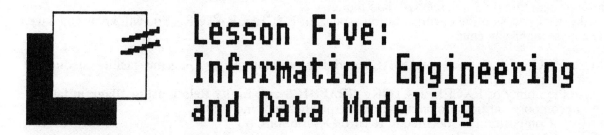

Lesson Five:
Information Engineering
and Data Modeling

The Demonstration Scenario

Soundstage Record Club's Data Processing Department is predominantly a COBOL shop. They also use IDMS, a Database Management System. The Systems Department has adopted a methodology that represents their interpretation of three popular methodologies:

    o   Structured Systems Analysis and Design (for all COBOL applications)
    o   Information Engineering (to assist with database aspects of all applications)
    o   Prototyping (for requirements determination and design of all user interfaces such as reports and screens)

    Information Engineering is a systems development methodology based a data modeling. New systems are engineering around stable but adaptable data models and procedures needed to maintain accurate data. Outputs are considered secondary since most immediate and future outputs can be delivered if essential data is captured and stored in flexible and adaptable subject databases. Data models become the foundation for database design and prototypes. At Soundstage, prototypes are built using ADS/O, a facility of IDMS. However, most applications are reprogrammed in COBOL/IDMS for final implementation.
    The requirements for the record club project will be defined around a third normal form data model. Facts about essential data have been collected. It is time to draw entity-relationship diagrams to depict the data model. Data records and elements must be cataloged into the data dictionary.

What Will You Learn in this Lesson?

In this lesson you will learn how to use EXCELERATOR to support documentation for *data modeling*. Data modeling is a technique associated with methodologies such as *Information Engineering*, as advocated by Martin and Finkelstein, and *Information Modeling*, as advocated by Flavin. Data models reflect the conceptual (implementation independent) structure of data that describes the system. Data models can, however, through an iterative process, evolve into implementation models for file and/or database management systems.
    The central tool in data modeling is the entity-relationship diagram (either the Chen or Merise notation) and the data model diagram (either the Bachman or Martin notation). All of these diagrams show business entities that are described by data elements and relationships between the business entities. Entities are ultimately decomposed into records that describe the elements that are contained in the entity. Ideally, the entities and data elements are *normalized*, a subject beyond the scope of this tutorial. The normalized data model serves as the basis for logical and physical database design or more creative and flexible conventional file design.

The decomposition of entities into records into data elements into codes is recorded into a data dictionary. The dictionary may also contain rules for updating and validating occurrences of entities. The graphs and data dictionary for data modeling can be developed directly from EXCELERATOR's **GRAPHICS** facility. This lesson will show you how to create *Information Engineering* specifications using EXCELERATOR. *Information Engineering* goes beyond pure data modeling to also define the procedures that are essential to maintain the implemented data model.

Specifically, we'll focus on the documentation aspects of EXCELERATOR. You will know you have mastered this exercise when you can:

1. Use EXCELERATOR's **GRAPHICS** and **Entity-Relationship Diagram** facility to draw a conceptual data model for a system.
2. Use the ability of EXCELERATOR's **GRAPHICS** and **Entity-Relationship Diagram** facility to decompose entities into data dictionary specifications including:
    o Composition as specified in **Record** Descriptions.
    o **Data Element** Descriptions.
    o Record updating events and conditions.

EXCELERATOR is also capable of checking the *Information Engineering* specification for errors and inconsistencies; however, that facility will be deferred until Lesson 7. Lesson 5 builds on the basic **GRAPHICS** skills that you learned in Lesson 3.

Our lesson uses the popular Entity-Relationship data modeling symbology developed by Peter Chen. EXCELERATOR also supports the Merise Entity-Relationship notation (if selected for your project account by the system or project manager) and the Bachman-like notation (accessed via the **Data Model Diagram** option on the Graphics Menu). The notations are *essentially* equivalent.

As we start the lesson, we just want to remind you that it is not our intent to teach or advocate *Information Engineering*. Some familiarity with the methodology or tools (or equivalents) would be helpful (a reference list is included at the end of this lesson). Also, be aware that every organization has different standards or conventions. Consequently, our approach may not precisely match yours. However, the intent of the lesson is to introduce the power of EXCELERATOR as a support tool for the basic concepts of data modeling.

Exercise 5.1   Draw an Entity-Relationship Diagram

Entity-relationship diagrams are one of the data modeling tools directly supported by EXCELERATOR. Although they can be exploded directly out of data stores on data flow diagrams, we will treat them separately. The separate entity-relationship diagram can be easily linked to a DFD by including its name in the explosion path for a data store (see Lesson 4). For this lesson, we want you to reproduce the entity-relationship diagram depicted in Figure 5-1.

Step 1.   Log on to EXCELERATOR and get to the Main Menu. SELECT **GRAPHICS** from the Main Menu. SELECT **Entity-Relationship Diagram** from the Graphics Menu. SELECT **Add** from the Action Keypad. For this graph, type the name *RECORD CLUB SYSTEM DATA MODEL*. Press the Enter key.

Step 2.   You should now see the drawing screen. Next, set the default drawing options. You did this in Lesson 3. For entity-relationship diagrams, these are as follows:

a.   SELECT **PROFILE** from the graphics commands. SELECT **GRID**. SELECT **FINE**. CANCEL to return to the Profile Menu. SELECT **CHGCONN**.

b.   CANCEL (remember, CANCEL always means click the right mouse button) to return to the commands menu. Your profile settings for this graph are complete.

c.   SELECT **PRINT**. Now SELECT **FULL GRAPH**. Your drawing quadrant is now visible. CANCEL the printing command.

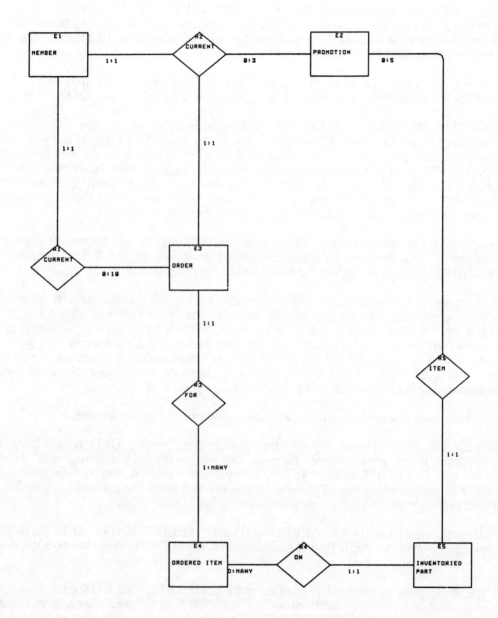

Figure 5-1

Entity-Relationship Diagram (Data Model) for the Record Club Project

75

Step 3.	Now you are ready to draw the entity-relationship diagram in Figure 5-1. You already know the basics. **SELECT OBJECT**. Notice the sub-menu below the graphics commands. There are only two object types. **SELECT ENTITY** from the object types. An entity is depicted as a box. **SELECT** locations for all of your entities. Next, **SELECT RELATION** from the object types. A relationship is depicted as a diamond. **SELECT** locations for the relationships. You can use the **MOVE** command to align the entities (as you learned in Lesson 3).

Next, add your connections. **SELECT CONNECT**. **SELECT** objects that are to be connected. An entity-relationship diagram's connections do not have arrows since the graph depicts structure, not flow. You can use the **MOVE** command to relocate the connections. This completes the basic graph.

Step 4.	Now we are ready to label the objects and connections. You may wish to **ZOOM** to **CLOSEUP** to see the labels better. If so, pan to the *MEMBER* entity in the drawing area.

SELECT DESCRIBE. **SELECT** the entity, *MEMBER*, on Figure 5-1. You are being prompted for an ID. Type *E1* and press the Enter key. This takes takes you into an Entity data dictionary description screen (Figure 5-2). The cursor is positioned in the **Label** area. Type *MEMBER*. If you wish, you can also move the cursor to the **Description** area and type a detailed description. Next, press the F3 function key. Recall that F3 means *save and exit*, which we'll use to exit all dictionary screens. This takes you back to the graph and the process has been named. Using Figure 5-1, name all entities in this fashion.

Step 5.	Next, name the relationships. Give them the IDs and labels shown in Figure 5-1. The relationship data dictionary description screen look like Figure 5-3. Remember, by using **DESCRIBE**, you are recording the relationships in the dictionary.

Step 6.	Next, name the connections. On entity-relationship diagrams, the our connection labels indicate *cardinality*, the complexity of the relationship. (This can also be recorded in the dictionary; however, we like to see the cardinalities on the graph.) For instance, in Figure 5-1, the label for the connections between *MEMBER* and *ORDER* suggests that for a *MEMBER*, there can be zero-to-many (0:10) *ORDER*s, but for an *ORDER*, there must be one and only one (1:1) *MEMBER*. If you are familiar with database techniques, then the importance of cardinality is equally familiar. The notation we suggest is read as follows:

minimum number of occurrences : maximum number of occurrences

Because the entity-relationship diagram connections, unlike DFD connections, contain no data, there is no need to enter them in the dictionary. Therefore, we'll use the **LABEL** command. **SELECT LABEL**. **SELECT** any connection. A pop-up window appears in the middle of the drawing area. Type the cardinality and press the Enter key. The label has been added to the graph. Add the remaining cardinality labels.

Step 7.	That completes Exercise 5.1. **SELECT OTHER**. **SELECT SAVE**. **SELECT EXIT** from the drawing commands. PRINTing is accomplished via the technique that you learned in Lesson 3.

If you don't plan to proceed immediately to Exercise 5.2, **SELECT Exit** from the Graphics Menu. Also, if you don't plan to proceed immediately to the next exercise, you might consider doing a **Backup** (via **HOUSEKEEPING**). Backup was also covered in Lesson 3, Exercise 3.3.

Exercise 5.2   Explode the Entities into Record Descriptions in the Data Dictionary

In *Information Engineering*, we *level* or *explode* entities into record descriptions that identify the data elements or attributes of that entity. Record descriptions are data structures that are recorded in a data dictionary. The notation used can take one of two forms: algebraic (as suggested by DeMarco and Figure 5-4)

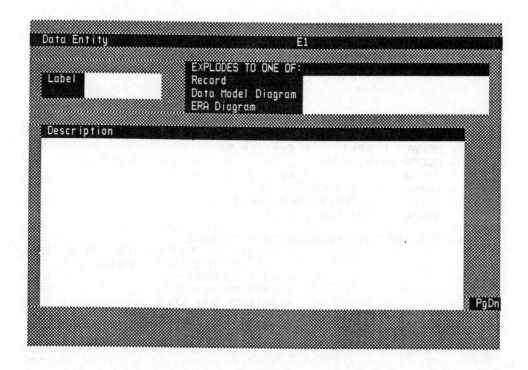

Figure 5-2

Data Entity Description Screen (in the Data Dictionary)

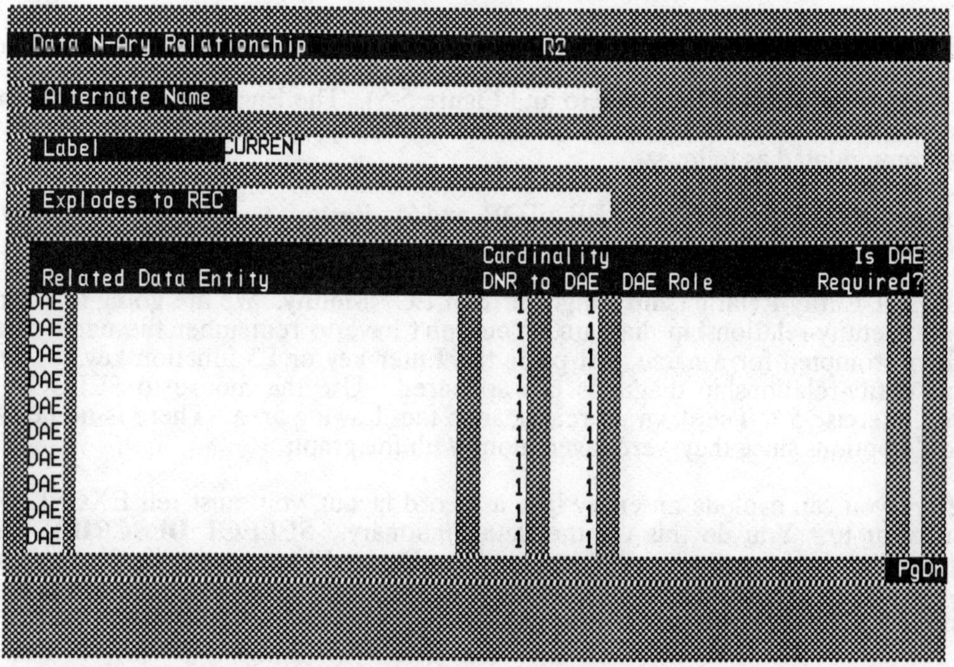

Figure 5-3

Relationship Description Screen (in the Data Dictionary)

```
ORDER  =  ORDER NUMBER        * primary key
        + MEMBER NUMBER        * foreign key
        + ORDER DATE
        + ORDER TYPE           * secondary key
        + (BANK CARD AUTHORIZATION)
        + (ALTERNATE SHIPPING ADDRESS:)
        + TOTAL AMOUNT DUE

ALTERNATE SHIPPING ADDRESS = SHIP TO NAME
                           + < SHIP TO POST OFFICE BOX NUMBER /
                               SHIP TO STREET ADDRESS >
                           + SHIP TO CITY
                           + SHIP TO STATE
                           + SHIP TO ZIP CODE
```

Figure 5-4

Algebraic Data Structure Notation (DeMarco)

or English (as suggested by Whitten/Bentley/Ho and Figure 5-5). The English notation more user friendly while the algebraic notation is more concise. EXCELERATOR supports neither notation; however, either notation can be accommodated as follows:

Step 1. If necessary log on to EXCELERATOR and/or **Restore** your project directory (Lesson 3). Proceed to the Graphics Menu.

Step 2. SELECT **Entity-Relationship Diagram.** SELECT **Modify.** We are going to explode *from* our context entity-relationship diagram. You don't have to remember the name of the diagram. When prompted for a name, just press the Enter key or F3 function key. A selector list of your entity-relationship diagrams has appeared. Use the mouse to SELECT the diagram from Exercise 5.1. The diagram reappear in the drawing area. There is no need to reset the profile options since they were saved along with the graph.

Step 3. Before you can explode an entity into a record layout, you must tell EXCELERATOR that you want to. You do this via the data dictionary. SELECT **DESCRIBE.** SELECT the *ORDER* entity from the drawing area. The ID should appear in the Status Line. Since we don't want to change the ID, press the Enter key. This takes you back to the data dictionary entry for the entity (Figure 5-6).

Notice that there is an **Explodes To One Of** area for describing what the entity explodes to. This defines the *explosion path* for the entity. We want to explode the entity to a **Data Record**; therefore, SELECT that line, the first blank position. Here, we must enter the name of the record to which we will explode. For entities, we prefer to give them the identical name of the entity from which they are exploded. For instance, we would explode the *ORDER* entity into the *ORDER* record layout. Don't worry! EXCELERATOR won't get confused. It identifies objects by a combination of object type (e.g., record) and ID or name

78

An occurrence of ORDER consists of the following:
    <u>ORDER NUMBER</u>
    MEMBER NUMBER as a pointer back to the MEMBER entity
    ORDER DATE
    ORDER TYPE as a sorting or grouping field
    BANK CARD AUTHORIZATION (optional)
    ALTERNATVE SHIPPING ADDRESS (optional) which consists of:
        SHIP TO NAME
        One or both of:
            SHIP TO POST OFFICE BOX NUMBER
            SHIP TO STREET ADDRESS
        SHIP TO CITY
        SHIP TO STATE
        SHIP TO ZIP CODE
    TOTAL AMOUNT DUE

Figure 5-5

English Data Structure Notation (Whitten/Bentley/Ho)

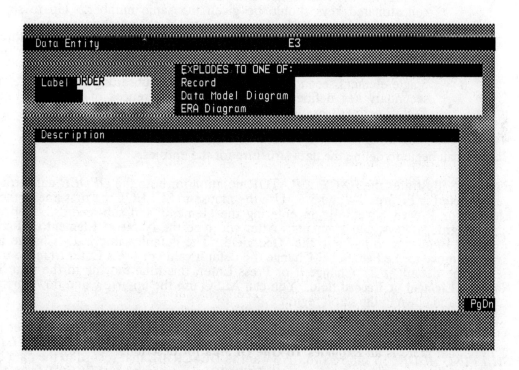

Figure 5-6

Data Entity Description Screen (in the Data Dictionary)

(e.g., *ORDER*). The other occurrence of *ORDER* has a different type, entity. Type the name of the record (entity). Press the F3 function key to save and exit the dictionary. You have returned to the graph. **DESCRIBE** all entities similarly.

Step 4. Now we can explode the entities. **SELECT EXPLODE**. SELECT the *ORDER* entity from the graph. A blank Record Description Screen (Figure 5-7) will appear. The sections of this screen are described as follows:

o **Alternate Name.** This is used to record an alias.

o **Definition.** This is a one-line description field.

o **Name of Element or Record.** This is used to record the names of elements or groups of elements (also known as a sub-record). This is a scroll region that can accommodate up to 115 entries.

o **Occ.** This stands for *number of occurrences of the element or sub-record in a single occurrence of the record*. Because most data models are normalized, this field will normally have the value 1.

o **Seq.** This stands for *sequence*. The numbers you place here determine the sequence of the elements and sub-records in the record. Thus, you can insert new elements or sub-records at the end of the record and alter the sequence such that they will print in the desired sequence.

o **Type.** This field is used to associate a type with the corresponding data element or sub-record. We have adopted the following internal legend:

E	Non-key data element
R	Sub-record: group of elements that always occur together
K	Single-element primary key
1-6	Single-element primary key or data element that is part of a concatonated key. All parts of the concatonated key should be given the same number. Alternate concatonated keys should be given the same number. Up to six keys and concatonated keys can be accommodated by the number scheme.
7	Single-element foreign key. A foreign key is a data element that links back to another entity's primary key (but is not part of its own record's primary or concatonated key).
8	Single-element secondary key (also called a subsetting criteria). A value for a secondary key defines a subset of all occurrences of the record (e.g., SEX CODE defines two subsets: male and female).
9	An optional, non-key data element

Now we can begin to define the data structure for the entities.

Step 5. Figure 5-8 illustrates the EXCELERATOR notation for both the *ORDER* data structure that was depicted in Figures 5-4 and 5-5. Use the mouse to SELECT the first line in the **Name of Element or Record** block. Begin entering the elements and sub-records. Notice how we handled groups, keys, and elements. After you type the **Name of Element or Record** field, press the Enter key to move to the **Occ.** field. The default value is *1*. Change it or press Enter to move to the **Seq.** field. Change the default value or press Enter to move to the **Type** field. The default is *E*. Change it or Press Enter, this time moving to the next line in the **Name of Element or Record** field. You can ALSO use the up-arrow and down-arrow keys to move up and down in the scroll region.

Step 6. You may be wondering how to define the data elements that make up the sub-records (Type = R). It's very easy. Use the mouse to position the cursor somewhere in the *ALTERNATE SHIPPING ADDRESS* sub-record. We want to **DESCRIBE** that record in the data dictionary. Unlike the graphics facility, there is no **DESCRIBE** command to SELECT from the screen. But there is a **DESCRIBE** key, the F4 function key. Press the F4 function key.

This takes you to a new Record Description screen, specifically for the ALTERNATE SHIPPING ADDRESS. Complete the layout per Figure 5-9.

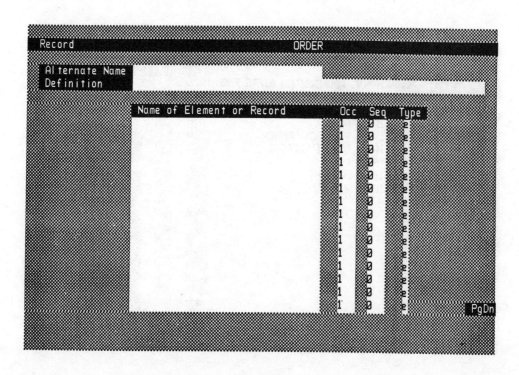

Figure 5-7

Data Record Description Screen (in the Data Dictionary)

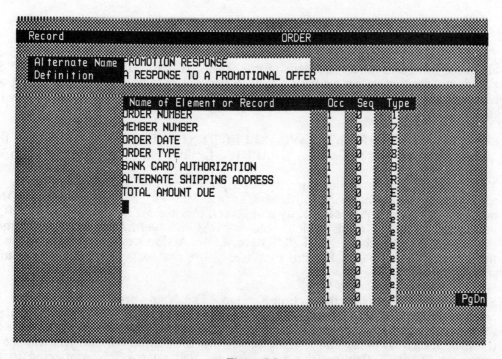

Figure 5-8

Completed Data Record Description Screen

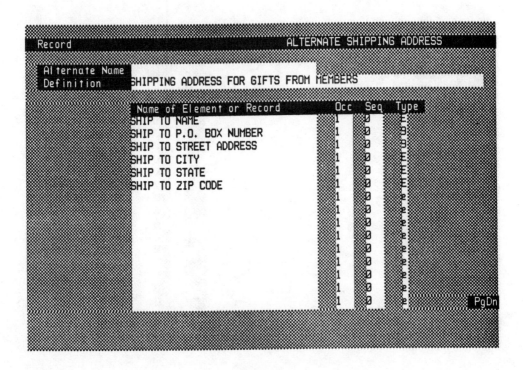

Figure 5-9

Record Description Screen for a Sub-Record or Group

Step 7. The rest of the record layouts for the data model are illustrated in Figure 5-10, some in algebraic form and others in English form. Convert them to EXCELERATOR descriptions. For demonstration of concatonated keys, a concept common to data modeling and normalization, the record description for *ORDERED ITEM* is depicted in Figure 5-11 (you still should enter it).

SELECT **OTHER**. SELECT **SAVE**. SELECT **EXIT**. This should take you to the Graphics Menu. SELECT **Exit** to leave the **GRAPHICS** facility. Now you are at the Main Menu.

Step 8. You cannot print your data dictionary descriptions from the **GRAPHICS** facility. If you want a hard copy, do the following (otherwise, skip to Step 9). SELECT **XLDICTIONARY**. This takes you to an entirely new facility and screen (Figure 5-12). From this XLDICTIONARY menu, SELECT **REC/ELE** (which stands for *RECords and ELEments*). A sub-menu appears in the middle of the screen. SELECT **Record**. An Action Keypad appears on the right side of the screen (Figure 5-13). For this exercise, ignore that large sub-menu in the middle of the screen.

SELECT **OUTPUT** from the Action Keypad. The **Name** prompt will appear. We want all records; therefore, type an asterisk (the wildcard character) and press the Enter key. The screen is replaced with a list of all records (similar to Figure 5-14). SELECT the first item on the list, labeled *All entities on selector list*. Below the Action Keypad you will see your output options. SELECT **Printer** (this step assumes that your PC is connected to a supported printer). After your output prints, SELECT **Exit** to return to the Main Menu.

Step 9. That completes Exercise 5.2. If you don't plan to proceed immediately, you might consider doing a **Backup** (via **HOUSEKEEPING**). Backup was covered in Lesson 3.

```
MEMBER = MEMBER NUMBER      * primary key
       + MEMBER NAME:       * first alternate primary key
       + DATE ENROLLED
       + (BANK CARD 1 AUTHORIZATION)
       + (BANK CARD 2 AUTHORIZATION)
       + MEMBER ADDRESS:
       + MUSICAL PREFERENCE    * secondary key
       + MEDIUM PREFERENCE     * secondary key
       + MEMBER BALANCE DUE
       + ACCOUNT STATUS        * secondary key
       + MEMBERSHIP PURCHASE REQUIREMENT
       + MEMBERSHIP EXPIRATION DATE      * secondary key
       + MEMBER PURCHASES TOWARD REQUIREMENT
       + MEMBER PURCHASES BEYOND REQUIREMENT
       + BONUS CREDITS CLAIMED

MEMBER NAME = LAST NAME
            + FIRST NAME
            + (MIDDLE INITIAL)
            + (TITLE) * Mr., Mrs., Miss, Ms., etc.

MEMBER ADDRESS = (STREET ADDRESS)
               + (POST OFFICE BOX NUMBER)
               + CITY
               + STATE
               + ZIP CODE
               + AREA CODE
               + PHONE NUMBER

An occurrence of ORDERED ITEM consists of the following:
     The concatenated key:
          ORDER NUMBER
          CATALOG NUMBER
     QUANTITY ORDERED
     QUANTITY FILLED
     UNIT PRICE AT TIME OF ORDER
     BONUS COUPON ISSUED?
```

Figure 5-10(a)

Data Structures for Remaining Entities and Records in Data Model (Part 1)

An occurrence of INVENTORIED PART consists of the following:
<u>CATALOG NUMBER</u>
<u>SUPPLIER PART NUMBER</u> as an alternate primary key
PRODUCT CATEGORY as a sorting or grouping field
PART DESCRIPTION which consists of the following:
 ARTIST DESCRIPTION (required only for recording titles)
 One of the following: ?
  CATALOG TITLE for recordings
  CATALOG DESCRIPTION for all other merchandise
CURRENT PRICE
SPECIAL PRICE
SHIPPING COST
QUANTITY IN STOCK
QUANTITY ON ORDER FROM SUPPLIER

PROMOTION = PROMOTION ID   * primary key
  + PROMOTION RELEASE DATE
  + PROMOTION AUTOMATIC ORDER FILL DATE * secondary key
  + PROMOTION EARNS MEMBER CREDITS?
  + ORDER EARNS BONUS SELECTIONS?

Figure 5-10(b)

Data Structures for Remaining Entities and Records in Data Model (Part 2)

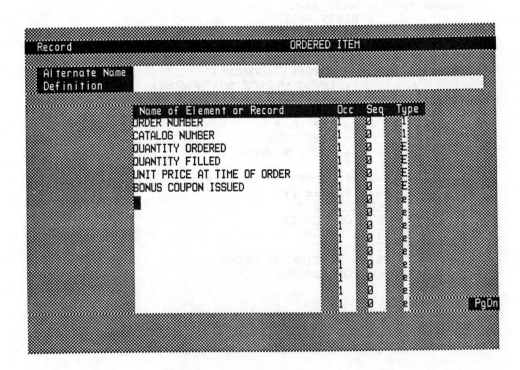

Figure 5-11

Entering Concatonated Keys Into a Data Record Description Screen

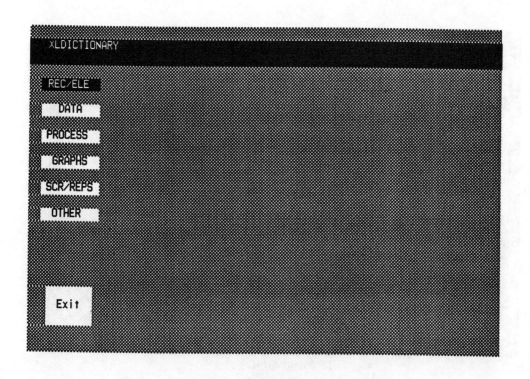

Figure 5-12

The XLDICTIONARY Submenu Screen

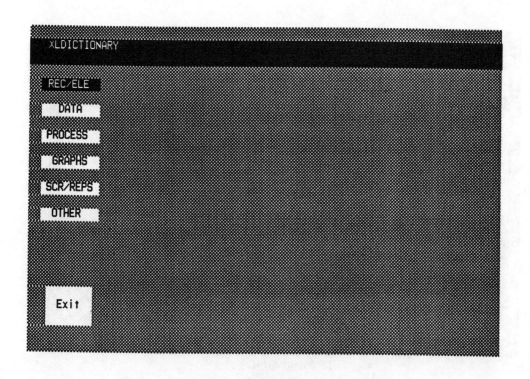

Figure 5-13

Action Keypad for XLDICTIONARY Submenu Screen

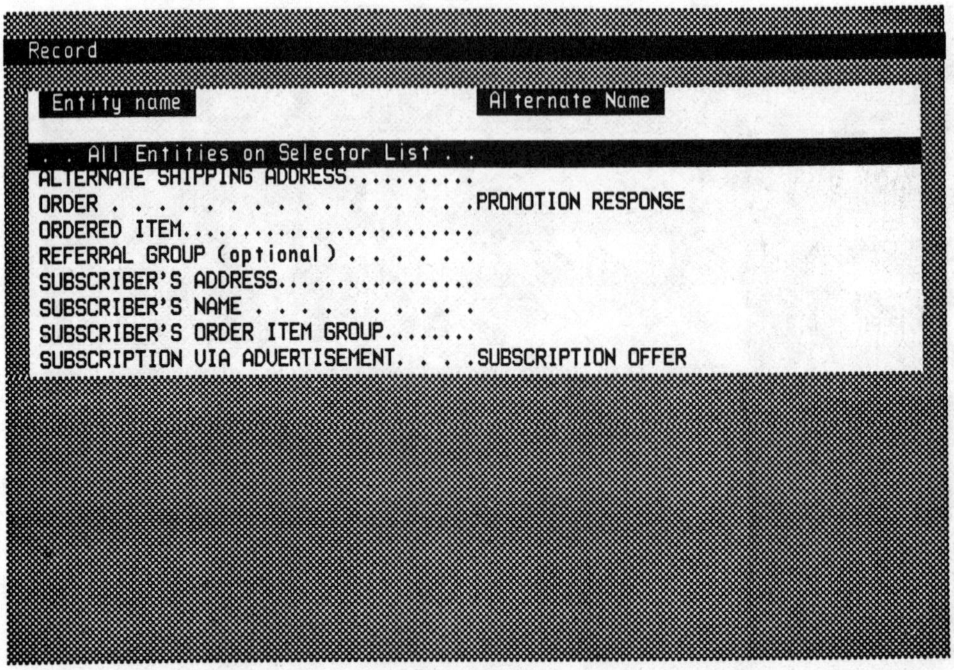

```
┌─────────────────────────────────────────────────────────────────────┐
│ Record                                                                │
│ ┌──────────────────────────┐        ┌──────────────────────────┐     │
│ │ Entity name              │        │ Alternate Name           │     │
│ └──────────────────────────┘        └──────────────────────────┘     │
│ ┌─────────────────────────────────────────────────────────────────┐  │
│ │ . . All Entities on Selector List . .                           │  │
│ │ ALTERNATE SHIPPING ADDRESS...........                           │  │
│ │ ORDER . . . . . . . . . . . . . . . .PROMOTION RESPONSE         │  │
│ │ ORDERED ITEM........................                            │  │
│ │ REFERRAL GROUP (optional) . . . . . .                          │  │
│ │ SUBSCRIBER'S ADDRESS.................                          │  │
│ │ SUBSCRIBER'S NAME . . . . . . . . .                            │  │
│ │ SUBSCRIBER'S ORDER ITEM GROUP........                          │  │
│ │ SUBSCRIPTION VIA ADVERTISEMENT. . . .SUBSCRIPTION OFFER         │  │
│ └─────────────────────────────────────────────────────────────────┘  │
│                                                                       │
└─────────────────────────────────────────────────────────────────────┘
```

Figure 5-14

Selector List for Data Records Described in the Data Dictionary

Exercise 5.3   Describe Data Elements to the Data Dictionary

Data modeling also requires that data elements be fully **DESCRIBE**d to the dictionary. This is particularly important since data elements, especially keys, are frequently associated with many entities. Thus, changes to the element dictionary can be automatically reflected in the rest of the specification. Specifying data elements in the dictionary can be executed via two facilities: **GRAPHICS** and **XLDICTIONARY**. We'll use the former since, most of the time, you want to define the data elements during the data modeling graphics stage of a project.

Step 1.   If necessary, log on to EXCELERATOR and/or **Restore** your project directory (Lesson 3). Proceed to the Graphics Menu. Call up your entity-relationship data model.

Step 2.   SELECT **EXPLODE** from the graphics commands. SELECT the *MEMBER* entity from the graph. The Record Description that you entered in the last exercise is now on the screen. SELECT the data element *MEMBER NUMBER*. Let's **DESCRIBE** the element to the dictionary.

Recall from the last exercise that the F4 function key is used to **DESCRIBE**. Press the F4 function key. This takes you to a new screen, a Data Element Description Screen (Figure 5-15). The components of this screen are described as follows:

*   **Alternate Names.** Up to three aliases can be entered. As a project moves closer to implementation, you could use alternate names to specify COBOL names, Data Definition Language names for a DBMS, etc.

86

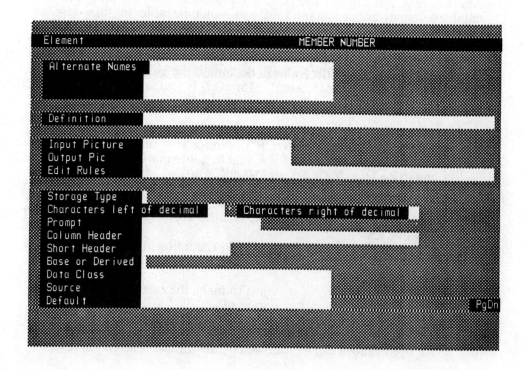

Figure 5-15

Data Element Description Screen (in the Data Dictionary)

* **Definition.** This field allows you to enter a brief, one-line description of the data element. Important: The entry in this field is used as a default help message in the screen prototyping facility (covered in Lesson 6).
* **Input Picture** and **Output Pic.** This field specifies the default input and output format for the data element. If you plan to use the screen and report design prototyping facilities, you should use EXCELERATOR's COBOL-like syntax so that you can take advantage of input editing and sample output generation capabilities. The input characters are:

A	for an alphabetic character
X	for an alphanumeric character
B	for a hard blank print position
S	for a positive or negative sign
9	for numeric print positions
.	for a period
,	for a comma

Output characters include all of the above plus:

$	for a dollar sign
+	for a plus sign
-	for a minus sign
z	for a non-zero fill blank

Alternatively, you can use the picture equivalents for your programming language or database management system (but you won't be able to fully test any prototypes developed in the Screens and Reports design facility).

* **Edit Rules.** As was the case with picture clauses above, you can use the **Edit Rules** field in two ways. If you plan to prototype or implement on a system other than EXCELERATOR, use **Edit Rules** to document the guidelines according to that system (e.g., PL/1, FOCUS, IDMS, etc.). However, if you want to use EXCELERATOR's pre-defined syntax:

o   For values that are alphanumeric strings, use double quotes (e.g. "*FRESH*", "*SOPH*", "*JUNIOR*", "*SENIOR*"). Notice that multiple values are separated by commas placed outside of the double quote marks. Wildcards are acceptable in any string (e.g., "*COLD**" or "**INSRT**")

o   For a finite series of numbers, do not use double quotes (e.g., *1,2,3,4*). Again, the values are separated by commas.

o   For a range of numeric values, use the keyword *THRU* (e.g., *4.25 THRU 12.35* as a pay scale).

o   All of the above can be preceded by the following selection rules:

NOT	
VALUE IS	just to make the rule easier to read
VALUES ARE	again, for readability

o   All of the above can be followed by the following selection rule:

OPT	to specify that a value does not have to be entered

o   You can also specify a table of codes from which values must or must not belong:

FROM "tablename"	The tablename must be in double quotes.
NOT FROM "tablename"	

The tablename may be followed by the *OPT* command to indicate that the element is optional. If you omit *OPT*, then the element must be input when requested on any screens you design using EXCELERATOR's screen design facility (Lesson 6).

You'll learn how to set up the table of codes in the next exercise.

* **Storage Type.** This is a one-character code whose values are as follows:

C	character
B	binary
P	packed
F	floating point
D	date

* **Characters left of decimal** and **Characters right of decimal.** This field defines the length of the data element. Only fill in characters to the right if **Storage Type is** *P* (*packed*).
* **Prompt.** This field establishes a default prompt that can be used in EXCELERATOR's screen and report design facilities (Lesson 6).
* **Column Header and Short Header.** These are used for documentation purposes only.
* **Base or Derived.** Is the element input (base) or calculated (derived)? The values are *B* or *D*. Since data models are usually normalized, the value is usually base

88

(calculated elements within a single record type are removed in third normal form). Derived elements are possible if they are derived from elements located in other records.

* **Data Class.** This field refers to a user assigned data grouping and is for documentation purposes.
* **Source.** This field refers to the source of information on this element.
* **Default.** If the element has a default value, it is recorded here.

Obviously, not all of the above fields will be used in every data modeling situation. All are available for use.

Step 3. Now that you understand the screen, make a few entries. The data elements for the *MEMBER* record are listed in Figure 5-16. Using that figure as a source of facts, complete the element specifications for the *MEMBER NUMBER*. Use the Enter key to progress through the entries. When you are finished with all the entries, press the F3 function key to save and exit.

You should now be back to the Record Description screen. SELECT the next element and press the F4 function key to **DESCRIBE** that element to the dictionary (use the data element dictionary entries appearing in Figure 5-16 as your source of facts). Repeat the process until all of the elements (not sub-records, Type *R*) are described.

After you have **DESCRIBE**d all the data elements for the *MEMBER* record, press the F3 function key to save and exit the Record Description screen. This takes you back to the entity-relationship diagram. SELECT **EXIT**. SELECT **Exit** again to return to the Main Menu.

Step 4. You cannot print your data dictionary descriptions from the **GRAPHICS** facility. If you want a hard copy, following a procedure similar to the printing of records in the last exercise. SELECT **XLDICTIONARY**. From the **XLDICTIONARY** menu, SELECT **REC/ELE**. A sub-menu appears in the middle of the screen. SELECT **ELEMENT**. An Action Keypad appears on the right side of the screen. SELECT **OUTPUT** from the Action Keypad. The **Name** prompt will appear. We want all data elements; therefore, type an asterisk (the wildcard character) and press the Enter key. The screen is replaced with a selector list of all data elements. SELECT the first item on the list, labeled *All entities on selector list*. Below the Action Keypad you will see your output options. SELECT **Printer** (this step assumes that your PC is connected to a supported printer). After your output prints, SELECT **Exit** to return to the Main Menu.

Step 5. That completes Exercise 5.3. If you don't plan to proceed immediately, you might consider doing a **Backup** (via **HOUSEKEEPING**). Backup was covered in Lesson 3.

TYPE Element   NAME ACCOUNT STATUS

   Alternate Names

   Definition   MEMBER'S ACCOUNT STATUS FOR FILLING ORDERS

   Input Picture   X
   Output Pic   X(10)
   Edit Rules   FROM "ACCOUNT STATUS CODES"

   Storage Type   C
   Characters left of decimal 1   Characters right of decimal 0
   Prompt   ACCOUNT STATUS
   Column Header   STATUS
   Short Header   STAT
   Base or Derived   B
   Data Class
   Source   ACCOUNTING
   Default   M

   Description

   Modified By   STUDENT ANALYST   Date Modified   861201   # Changes 0
   Added By   STUDENT ANALYST   Date Added   861201
   Last Project   COURSE PROJECT
   Locked By   Date Locked   0   Lock Status

TYPE Element   NAME BANK CARD 1 AUTHORIZATION

   Alternate Names

   Definition   BANK CARD NUMBER TO BE USED WHEN ORDER REQUESTS IT BE USED

   Input Picture   X999999999999999999999
   Output Pic   X999999999999999999999
   Edit Rules   OPT

   Storage Type   C
   Characters left of decimal 22   Characters right of decimal 0
   Prompt   BANKCARD NUMBER
   Column Header
   Short Header
   Base or Derived   B
   Data Class
   Source   APPLICATION
   Default

   Description

   Modified By   STUDENT ANALYST   Date Modified   861201   # Changes 0
   Added By   STUDENT ANALYST   Date Added   861201
   Last Project   COURSE PROJECT
   Locked By   Date Locked   0   Lock Status

Figure 5-16(a)

Data Element Descriptions for the Member Data Record (Part 1)

```
DATE: 11-MAR-87      ELEMENT - OUTPUT                          PAGE     1
TIME: 11:42          NAME: BANK CARD 2 AUTHORIZATION      EXCELERATOR  1.7

TYPE Element                          NAME BANK CARD 2 AUTHORIZATION
  Alternate Names


  Definition     BANK CARD NUMBER TO BE USED IN EVENT OF CREDIT REJECT ON #1

  Input Picture  X9999999999999999999999
  Output Pic     X9999999999999999999999
  Edit Rules     OPT

  Storage Type   C
  Characters left of decimal 22    Characters right of decimal 0
  Prompt         BANK CARD 2 NUMBER
  Column Header
  Short Header
  Base or Derived B
  Data Class
  Source         APPLICATION
  Default


  Description


  Modified By   STUDENT ANALYST    Date Modified  861201    # Changes  0
  Added By      STUDENT ANALYST    Date Added     861201
  Last Project  COURSE PROJECT
  Locked By                        Date Locked    0         Lock Status
```

```
DATE: 11-MAR-87      ELEMENT - OUTPUT                          PAGE     1
TIME: 11:49          NAME: BONUS CREDITS EARNED          EXCELERATOR  1.7

TYPE Element                          NAME BONUS CREDITS EARNED
  Alternate Names


  Definition     BONUS COUPONS THAT HAVE YET TO BE REDEEMED BY MEMBER

  Input Picture  99
  Output Pic     99
  Edit Rules     VALUES ARE 0 THRU 25

  Storage Type   P
  Characters left of decimal 2     Characters right of decimal 0
  Prompt
  Column Header  COUPONS
  Short Header   COUP
  Base or Derived D
  Data Class
  Source         ORDERS
  Default        0


  Description


  Modified By   STUDENT ANALYST    Date Modified  861201    # Changes  0
  Added By      STUDENT ANALYST    Date Added     861201
  Last Project  COURSE PROJECT
  Locked By                        Date Locked    0         Lock Status
```

Figure 5-16(b)

Data Element Descriptions for the Member Data Record (Part 2)

91

```
DATE: 11-MAR-87      ELEMENT - OUTPUT                          PAGE    1
TIME: 11:41          NAME: DATE ENROLLED              EXCELERATOR  1.7

TYPE Element                            NAME DATE ENROLLED

   Alternate Names


   Definition        DATE THAT MEMBER FIRST ENROLLED WITH RECORD CLUB

   Input Picture     999999
   Output Pic        99/99/99
   Edit Rules

   Storage Type      D
   Characters left of decimal 6        Characters right of decimal 0
   Prompt            DATE ENROLLED
   Column Header     DATE ENROLLED
   Short Header      DATE ENR
   Base or Derived   B
   Data Class
   Source            DATE STAMPED ON APPLICATION
   Default


   Description


   Modified By    STUDENT ANALYST    Date Modified   861201     # Changes   0
   Added By       STUDENT ANALYST    Date Added      861201
   Last Project   COURSE PROJECT
   Locked By                         Date Locked        0       Lock Status
```

```
DATE: 11-MAR-87      ELEMENT - OUTPUT                          PAGE    1
TIME: 11:46          NAME: MEDIUM PREFERENCE          EXCELERATOR  1.7

TYPE Element                          NAME MEDIUM PREFERENCE

   Alternate Names


   Definition        DEFAULT RECORDING MEDIUM TO FILL MEMBER'S ORDERS

   Input Picture     XX
   Output Pic        X(10)
   Edit Rules        FROM "MEDIA CODES"

   Storage Type      C
   Characters left of decimal 2        Characters right of decimal 0
   Prompt            MEDIUM
   Column Header     MEDIUM
   Short Header      MED
   Base or Derived   B
   Data Class
   Source            APPLICATION or PHONE CALL
   Default           RC


   Description


   Modified By    STUDENT ANALYST    Date Modified   870309     # Changes   1
   Added By       STUDENT ANALYST    Date Added      861201
   Last Project   COURSE PROJECT
   Locked By                         Date Locked        0       Lock Status
```

Figure 5-16(c)

Data Element Descriptions for the Member Data Record (Part 3)

92

TYPE Element   NAME MEMBER BALANCE DUE

   Alternate Names

   Definition   UNPAID BALANCE DUE ON MEMBER'S ACCOUNT

   Input Picture   999.99
   Output Pic   -/+$ZZ9.99
   Edit Rules   VALUES ARE -50.00 THRU 150.00

   Storage Type   P
   Characters left of decimal 3   Characters right of decimal 2
   Prompt   NEW BALANCE
   Column Header   BALANCE
   Short Header
   Base or Derived   D
   Data Class
   Source   ACCOUNTING
   Default   0.00

   Description

   Modified By   STUDENT ANALYST   Date Modified 861201   # Changes 0
   Added By   STUDENT ANALYST   Date Added   861201
   Last Project   COURSE PROJECT
   Locked By   Date Locked   0   Lock Status

TYPE Element   NAME MEMBER NAME

   Alternate Names

   Definition   LAST NAME, FIRST NAME MIDDLE INITIAL

   Input Picture   X(30)
   Output Pic   X(30)
   Edit Rules

   Storage Type   C
   Characters left of decimal 30   Characters right of decimal 0
   Prompt   NAME:LAST, FIRST INT
   Column Header   MEMBER NAME
   Short Header
   Base or Derived   B
   Data Class
   Source   APPLICATION
   Default

   Description

   Modified By   STUDENT ANALYST   Date Modified 861201   # Changes 1
   Added By   STUDENT ANALYST   Date Added   861201
   Last Project   COURSE PROJECT
   Locked By   Date Locked   0   Lock Status

Figure 5-16(d)

Data Element Descriptions for the Member Data Record (Part 4)

TYPE Element   NAME MEMBER NUMBER

   Alternate Names MEMBERSHIP NUMBER

   Definition   A 7 DIGIT NUMBER THAT UNIQUELY IDENTIFIES A MEMBER

   Input Picture   9999999
   Output Pic   9999999
   Edit Rules   0000001 THRU 9999999

   Storage Type   C
   Characters left of decimal 7   Characters right of decimal 0
   Prompt   MEMBERSHIP NUMBER?
   Column Header   MEMBER NUMBER
   Short Header   MEMBER#
   Base or Derived D
   Data Class
   Source   SUBSCRIPTION MANAGER
   Default

   Description

   Modified By   STUDENT ANALYST   Date Modified   870309   # Changes 18
   Added By   STUDENT ANALYST   Date Added   861111
   Last Project   COURSE PROJECT
   Locked By   Date Locked   0   Lock Status

TYPE Element   NAME MEMBERSHIP EXPIRATION DATE

   Alternate Names

   Definition   DATE BY WHICH MINIMUM REQUIRED PURCHASES MUST BE MADE

   Input Picture   9999
   Output Pic   99/99
   Edit Rules   MMYY

   Storage Type   D
   Characters left of decimal 4   Characters right of decimal 0
   Prompt   EXPIRATION DATE
   Column Header   EXPIRE DATE
   Short Header   EXPIRE
   Base or Derived B
   Data Class
   Source   BASED ON APPLICATION
   Default

   Description

   Modified By   STUDENT ANALYST   Date Modified   861201   # Changes 0
   Added By   STUDENT ANALYST   Date Added   861201
   Last Project   COURSE PROJECT
   Locked By   Date Locked   0   Lock Status

Figure 5-16(e)

Data Element Descriptions for the Member Data Record (Part 5)

```
DATE: 11-MAR-87        ELEMENT - OUTPUT                                PAGE     1
TIME: 11:48            NAME: MEMBER PURCHASES BEYOND REQT        EXCELERATOR  1.7

TYPE Element                              NAME MEMBER PURCHASES BEYOND REQT

   Alternate Names

   Definition         PURCHASES BEYOND MINIMUM AGREEMENT, COUNTING FOR BONUS CRDTS

   Input Picture      999
   Output Pic         999
   Edit Rules         VALUES ARE 0 THRU 999

   Storage Type       P
   Characters left    of decimal 3      Characters right of decimal 0
   Prompt
   Column Header      BONUS PURCH
   Short Header       BON
   Base or Derived    B
   Data Class
   Source             ORDERS
   Default            0

   Description

   Modified By      STUDENT ANALYST    Date Modified    861201      # Changes   0
   Added By         STUDENT ANALYST    Date Added       861201
   Last Project     COURSE PROJECT
   Locked By                           Date Locked      0           Lock Status
```

```
DATE: 11-MAR-87        ELEMENT - OUTPUT                                PAGE     1
TIME: 11:47            NAME: MEMBERSHIP PURCHASE REQUIREMENT     EXCELERATOR  1.7

TYPE Element                              NAME MEMBERSHIP PURCHASE REQUIREMENT
   Alternate Names

   Definition         THE NUMBER OF PURCHASES MEMBER MUST MAKE IN SPECIFIED TIME

   Input Picture      99
   Output Pic         99
   Edit Rules         VALUES ARE 1,2,3,4,5,6,7,8,9,10,11,12,13,14,15

   Storage Type       P
   Characters left    of decimal 2      Characters right of decimal 0
   Prompt             PURCHASE AGREEMENT
   Column Header      REQUIRED
   Short Header       REQ
   Base or Derived    B
   Data Class
   Source             BASED ON APPLICATION
   Default

   Description

   Modified By      STUDENT ANALYST    Date Modified    861201      # Changes   0
   Added By         STUDENT ANALYST    Date Added       861201
   Last Project     COURSE PROJECT
   Locked By                           Date Locked      0           Lock Status
```

Figure 5-16(f)

Data Element Descriptions for the Member Data Record (Part 6)

```
DATE: 11-MAR-87        ELEMENT - OUTPUT                                    PAGE     1
TIME: 11:46            NAME: MUSICAL PREFERENCE                      EXCELERATOR  1.7

TYPE Element                                  NAME MUSICAL PREFERENCE

  Alternate Names

   Definition         CATEGORY OF MUSIC PREFERRED BY MEMBER.

   Input Picture      XX
   Output Pic         X(16)
   Edit Rules         FROM "MUSIC CATEGORY"

   Storage Type       C
   Characters left of decimal 2     Characters right of decimal 0
   Prompt             MUSICAL PREFERENCE
   Column Header      MUSIC PREF.
   Short Header       MUSIC
   Base or Derived B
   Data Class
   Source             APPLICATION
   Default

   Description

   Modified By     STUDENT ANALYST    Date Modified    870311      # Changes   1
   Added By        STUDENT ANALYST    Date Added       870311
   Last Project    COURSE PROJECT
   Locked By                          Date Locked      0           Lock Status
```

```
DATE: 11-MAR-87        ELEMENT - OUTPUT                                    PAGE     1
TIME: 11:48            NAME: MEMBER PURCHASE CREDITS EARNED          EXCELERATOR  1.7

TYPE Element                                  NAME MEMBER PURCHASE CREDITS EARNED

  Alternate Names

   Definition         NUMBER OF PURCHASES THAT COUNT TOWARD FULFILLING AGREEMENT

   Input Picture      99
   Output Pic         99
   Edit Rules         VALUES ARE 0,1,2,3,4,5,6,7,8,9,10,11,12,13,14,15

   Storage Type       P
   Characters left of decimal 2     Characters right of decimal 0
   Prompt
   Column Header      CREDITS
   Short Header       CRDT
   Base or Derived D
   Data Class
   Source             ORDERS
   Default            0

   Description

   Modified By     STUDENT ANALYST    Date Modified    861201      # Changes   1
   Added By        STUDENT ANALYST    Date Added       861201
   Last Project    COURSE PROJECT
   Locked By                          Date Locked      0           Lock Status
```

Figure 5-16(g)

Data Element Descriptions for the Member Data Record (Part 7)

Exercise 5.4   Define a Code Table for a Data Element

Code tables define legal (or illegal) business codes for data elements. They are needed to prototype and implement appropriate editing for the data elements. In EXCELERATOR code tables can be easily created and used for multiple data elements. They are associated with data elements through the **Edit Rules** entry in the data element dictionary (last exercise). Code tables can and should be shared between projects.

Step 1.   From EXCELERATOR's Main Menu, SELECT **XLDICTIONARY**. This takes you to the Data Dictionary Menu. SELECT **DATA**. SELECT **Table of Codes**.

For practice, we want to establish a table of codes for the *MUSICAL PREFERENCE* data element that was created in the last exercise. The **Edit Rules** for that element were *FROM "MUSIC CATEGORY"*. Therefore, we'll create the table *MUSICAL CATEGORY* (shown in Figure 5-17).

From the Action Keypad, SELECT **Add**. In the name field, type *MUSIC CATEGORY*. Press the Enter key. This takes you to a blank Table of Codes Screen. You can enter an **Alternate Name** and/or **Definition**. SELECT the first line in the **Code** column. Type the first code. Press the Enter key to get to the **Meaning** column. Type the meaning of the code. Press the Enter key to proceed to the next code. After all the codes and meanings are entered, press the F3 function key to save and exit. This returns you to the Data Dictionary Menu.

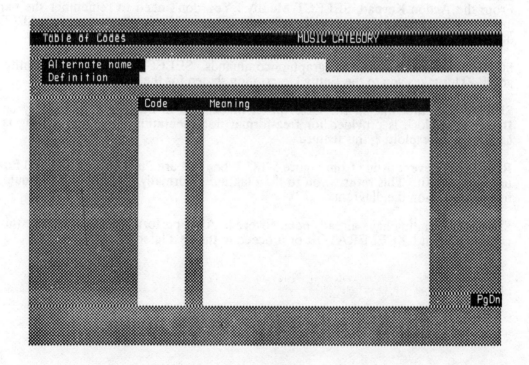

Figure 5-17

Table of Codes Description Screen

97

Step 2. If you want a printed copy of your code table, do the following (otherwise, skip to Step 3). From the Data Dictionary Menu, SELECT **Output** from the Action Keypad. The **Name** prompt will appear. Type *MUSIC CATEGORY* and press the Enter key. Below the Action Keypad you will see your output options. SELECT **Printer** (this step assumes that your PC is connected to a supported printer). After your output prints, SELECT **Exit** to return to the Main Menu.

Step 3. That's it for this lesson. If you don't plan to proceed immediately, you might consider doing a **Backup** (via **HOUSEKEEPING**). Backup was covered in Lesson 3.

Exercise 5.5 Describe Update Events and Conditions for a Data Entity

A very important aspect of data modeling is event modeling. For each entity, you identify and record those business events that:

o create occurrences of the entity
o delete occurrences of the entity
o modify occurrences of the entity

Events represent logic that will eventually be programmed to maintain the database. We can add these events to our dictionary, through either **GRAPHICS** or **XLDICTIONARY**. We recommend **GRAPHICS** since that gives you the ability to return to the entity-relationship diagram via the explosion path. Here's how it's done:

Step 1. From the Main Menu, SELECT **GRAPHICS**. Then, SELECT **Entity-Relationship Diagram**. From the Action Keypad, SELECT **Modify**. You don't need to remember the name. Just press the Enter key. A selector list of entity-relationship diagrams appears. SELECT your diagram from the list. The diagram should reappear.

Step 2. SELECT **EXPLODE** from the graphics commands. SELECT the *MEMBER* Entity from the graph. This takes you to the Entity Description screen for the *MEMBER* entity.

Step 3. SELECT the first line in the **Description** block. As in other data dictionary entries, the **Description** block is provided for free-format documentation. Event modeling is an ideal candidate for exploiting this feature.

Recreate the event listing from Figure 5-18. When you are finished, press the F3 function key to save and exit. This returns you to the diagram. Normally, a similar entry would be made for all entities on the diagram.

Step 4. Saving and printing have already been covered. After performing those actions, this lesson is complete. Exit EXCELERATOR or proceed to the next lesson.

```
DATE: 11-MAR-87        DATA ENTITY - OUTPUT                            PAGE    1
TIME: 11:52            NAME: E1                              EXCELERATOR  1.7


TYPE Data Entity                                NAME E1

                                       EXPLODES TO ONE OF:
        Label MEMBER                   Record                  MEMBER
                                       Data Model Diagram
                                       ERA Diagram


   Description
   Transactions that affect the MEMBER entity:
      Occurrences of MEMBER are created when:
          1. SUBSCRIPTIONs are received (advertisement or referral)
          2. Past member reactivates account (if ACCOUNT STATUS = I and
                                                 ACCOUNT BALANCE < 10.00)
      Occurrences of MEMBER are deleted (archived) when
          1. Member requests termination (if MEMBER PURCHASE CREDITS EARNED
                                            greater than or equal to MEMBER
                                                    PURCHASE REQUIREMENT)
          2. Accounts Receivable Dept changes ACCOUNT STATUS = D
          3. Order Dept changes ACCOUNT STATUS = I
      Occurrences of MEMBER can change under the following circumstances
      (grouped by data elements):
          1. For MEMBER NAME, BANK CARD n AUTHORIZATION, MEMBER ADDRESS,
             MUSICAL PREFERENCE, and MEDIUM PREFERENCE when requested by
             the member (usually via phone or correspondence)
          2. For MEMBER BALANCE
             a. When ORDER received
             b. When PAYMENT received
             c. When Return for Credit approved
          3. For ACCOUNT STATUS
             a. When Past Due Accounts frozen by Accounts Receivable Dept.
             b. When customer terminate Account
             c. When A/R terminates a delinquent account
             d. When account is inactive for 2 years
          4. For MEMBER PURCHASE CREDITS EARNED and MEMBER PURCHASE BONUS
             REQUIREMENTS
             a. When ORDER received
             b. When Return for Credit approved
          5. For BONUS CREDITS EARNED
             a. When Order received (if MEMBER PURCHASE CREDITS EARNED >
                MEMBER PURCHASE REQUIREMENT or ORDER TYPE = B)
             b. When Bonus Order received


Modified By     STUDENT ANALYST    Date Modified   870311      # Changes   8
Added By        STUDENT ANALYST    Date Added      861123
Last Project    COURSE PROJECT
Locked By                          Date Locked     0           Lock Status
```

Figure 5-18

Data Entity Description With Event Analysis

Conclusion

Data modeling methodologies are easily supported by EXCELERATOR. First, you draw the Entity-Relationship Diagram via the **GRAPHICS** facility. Entities on that graph are exploded into records. Records are **DESCRIBE**d in terms of their third normal form data elements. Data elements are **DESCRIBE**d in terms of their properties. Code tables can be created for data elements via the **XLDICTIONARY** facility. They are associated with data elements via the **Editing Rules** in the data element dictionary form.

Looking ahead to design, you can make copies of the ERDs, deleting entities and relationships in order to define sub-schemas. You can then draw a high-level data flow diagram (Lessons 3 and 4) that contains a single data store that explodes to the full-blown ERD. As the data flow diagrams are leveled, new data stores are exploded to the sub-schema ERDs. Consequently, there are opportunities to jointly use data modeling methodologies concurrently with data flow methodologies.

Information Engineering, like *Structured Analysis*, is a prespecification technique. This means that the methodology produces a specification document as a preface to building the system. In the next lesson, you'll learn how to *prototype* portions of a system using EXCELERATOR. Prototypes are used to help users see working subsets of a proposed system. This often helps users more accurately assess the analysts' interpretation of system requirements.

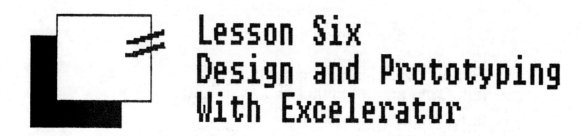

Lesson Six
Design and Prototyping
With Excelerator

The Demonstration Scenario

Soundstage Record Club's Data Processing Department is predominantly a COBOL shop. They also use IDMS, a Database Management System. The Systems Department has adopted a methodology that represents their interpretation of three popular methodologies:

    o   Structured Systems Analysis and Design (for all COBOL applications)
    o   Information Engineering (to assist with database aspects of all applications)
    o   Prototyping (for requirements determination and design of all user interfaces such as reports and screens)

    Prototyping at Soundstage takes on two forms. The IDMS database environment includes ADS/O, a facility for generating full scale prototypes of systems. Most prototypes are discarded are reprogrammed in COBOL/IDMS for maximum operating throughput and efficiency. However, long before full-scale prototypes are developed, record club analysts have learned that rapid prototypes of reports and screens prove very useful for getting agreement on user requirements and user interfaces - before any significant programming effort is needed.
    You are currently defining user requirements for the Customer Services System project. You are developing a database model (Lesson 5) for the new system. You have also started drawing and reviewing data flow diagrams for the new system (Lesson 4). You are having difficulty getting users to state and/or agree on detailed requirements for a few reports and inputs (which currently appear in your specifications as data flows on DFDs). Consequently, you have decided to generate rapid prototypes of these reports and screens in the hope that the users "will recognize their needs and requirements when they see them in operation."

What Will You Learn in this Lesson?

In this lesson you will learn how to use EXCELERATOR to support documentation, design, and prototyping of screens and reports. *Prototyping* is a systems development strategy that is being used by an increasing number of businesses. Through prototyping, you quickly develop working models of proposed systems, or portions thereof. The model is commonly referred to as a *prototype*. The prototype system is reviewed by users and management, who then make recommendations about requirements, methods, and formats. The prototype is then corrected, enhanced, or refined to reflect the new requirements. This process continues until the prototype evolves into the final design.
    There are two generally recognized prototyping approaches, *rapid prototyping* and *systems prototyping*. Rapid prototyping is limited to the modeling or design of the system's outputs, inputs, dialogues, and files. Systems prototyping is more encompassing. Systems prototyping also includes design and construction of a total working system including database/files, methods, and procedures.

The trend toward total systems prototyping has been fueled by the availability of special purpose software called *application generators*, *program code generators*, and *fourth-generation languages*. EXCELERATOR does not include such languages. However, through their add-on product, *CUSTOMIZER*, such tools can be integrated into the EXCELERATOR environment. EXCELERATOR does, on the other hand, include a number of design and rapid prototyping tools that work very nicely with their data dictionary.

In this lesson you will learn how to use EXCELERATOR for design and rapid prototyping of screens and reports. You will also learn about the limited code generation capabilities of EXCELERATOR. You will know you have mastered this lesson when you can:

1. Use EXCELERATOR's **SCREENS & REPORTS** and **Report Design** facility to design and prototype printed reports.
2. Use EXCELERATOR's **SCREENS & REPORTS** and **Screen Design** facility to plan and develop prototypes of on-line input, output, and dialogue screens.
3. Use EXCELERATOR's **SCREENS & REPORTS** and **Screen Data Entry** facility to test screen prototypes using actual data and to create screen test data files.
4. Use EXCELERATOR's **SCREENS & REPORTS** and **Screen Data Reporting** facility to generate program code of record layout descriptions for screens.

Exercise 6.1   Design a Report

EXCELERATOR's **SCREENS & REPORTS** and **Report Design** facility aids in the development of printed report designs and prototypes. Report designs and prototypes developed using EXCELERATOR may exhibit many of the format characteristics common to most reports (such as report and column headers, detail and summary lines, etc.).

Essentially the report you will design and prototype identifies the record club's order responses received during a given day. The report is a *group indicated* (multiple-level control break) report sequenced by member within musical preference category.

A sample design for our report is depicted in Figure 6-1. The design may look familiar to you, since it is typical of report designs you may have seen or documented using printer spacing charts.

Although there is a trend away from developing report designs in favor of report prototypes, report designs are still very useful. Report designs are, for instance, appropriate for communicating report format requirements to computer programmers. Let's learn how to use EXCELERATOR to document the report design in Figure 6-1.

Step 1.   Log on to EXCELERATOR and get to the Main Menu. If you haven't already done so, you need to Restore the work you've completed in Exercises 3 through 5. SELECT **SCREENS & REPORTS** from the Main Menu. A sub-menu appears (Figure 6-2). Now SELECT **Report Design** from the sub-menu. An Action Keypad should appear. SELECT **Add**. For this report, type the name *ORDER RESPONSE REPORT DESIGN*. Press the Enter key.

Step 2.   A blank Report Design Description Screen should appear (Figure 6-3). Let's add the following **Description:** *THIS REPORT DETAILS THE ORDER RESPONSES RECEIVED DURING A SINGLE DAY. THE REPORT IS GROUP INDICATED TO DIFFERENTIATE ORDERS BY MEMBER AND MUSICAL PREFERENCE CATEGORY.* The **Date Created:** and **Date Last Modified:** fields are maintained by EXCELERATOR. SELECT the **Created By:** entry. Type your name.

The last four items enable us to customize the size of the report (and, subsequently, the drawing screen). The default defines a report that consists of 66 lines of 132 columns each. These defaults can be changed to accommodate the type of printer and size of paper you are using. We will use the current default values for this exercise. Press the F3 function key to save, exit, and proceed to the Report Design Drawing Screen (Figure 6-4).

MUSICAL PREFERENCE	SELECTION OF MONTH TITLE	MEMBERSHIP NUMBER	SELECTION OF MONTH DECISION	CREDITS EARNED	CATALOG NUMBER	MEDIA	MESSAGES
XXXXXXXXXXXXXXX	XXXXXXXXXXXXXXXXXXX	99999	XXXX	Z9	99999 99999 99999	XXXXXXXXXXXXXXX XXXXXXXXXXXXXXX XXXXXXXXXXXXXXX	XXXXXXXXXXXXXXX
		99999	XXXX	Z9	99999	XXXXXXXXXXXXXXX	XXXXXXXXXXXXXXX
		99999	XXXX	Z9			XXXXXXXXXXXXXXX
	TOTAL ACCEPTING MONTHLY TITLE	ZZ9					
	TOTAL REJECTING MONTHLY TITLE	ZZ9					
XXXXXXXXXXXXXXX	XXXXXXXXXXXXXXXXXXX	99999	XXXX	Z9	99999	XXXXXXXXXXXXXXX	XXXXXXXXXXXXXXX
		99999	XXXX	Z9			XXXXXXXXXXXXXXX
		99999	XXXX	Z9	99999 99999 99999	XXXXXXXXXXXXXXX XXXXXXXXXXXXXXX XXXXXXXXXXXXXXX	XXXXXXXXXXXXXXX
	TOTAL ACCEPTING MONTHLY TITLE	ZZ9					
	TOTAL REJECTING MONTHLY TITLE	ZZ9					
XXXXXXXXXXXXXXX	XXXXXXXXXXXXXXXXXXX	99999	XXXX	Z9	99999	XXXXXXXXXXXXXXX	XXXXXXXXXXXXXXX
		99999	XXXX	Z9	99999	XXXXXXXXXXXXXXX	XXXXXXXXXXXXXXX
		99999	XXXX	Z9			XXXXXXXXXXXXXXX
		99999	XXXX	Z9	99999 99999	XXXXXXXXXXXXXXX XXXXXXXXXXXXXXX	XXXXXXXXXXXXXXX
	TOTAL ACCEPTING MONTHLY TITLE	ZZ9					
	TOTAL REJECTING MONTHLY TITLE	ZZ9					

NUMBER AND PERCENT OF RESPONSES FOR ORDER   ZZ9   Z9%
NUMBER AND PERCENT OF RESPONSES FOR NO ORDER   ZZ9   Z9%
NUMBER OF AUTOMATIC ORDERS (NONRESPONDENTS)   ZZ9

TOTAL NUMBER OF ORDERS   Z,ZZ9

Figure 6-1

Order Response Report Design for the Record Club Project

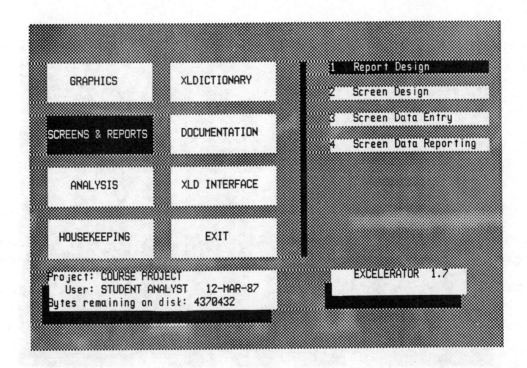

Figure 6-2

SCREENS & REPORTS Submenu Screen

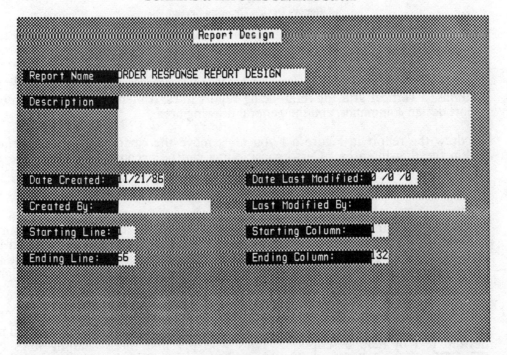

Figure 6-3

Report Design Description Screen

Figure 6-4

Report Design Drawing Screen

Step 3.   The Report Design Drawing Screen consists of a horizontal grid for referencing report columns, a vertical grid for referencing report lines, a menu (bottom of screen) containing report design commands, and the general drawing area.

To draw the report in Figure 6-1 you must move the cursor to the appropriate location(s) within the drawing area. Cursor positioning can be accomplished using the mouse or the arrow keys. The mouse can only SELECT positions on the screen. Practice using the mouse to SELECT positions.

Notice that the drawing screen area does not represent the entire size of our report. To quickly position the cursor to parts of the screen drawing area that are not currently visible, you must use two key strokes (not the mouse), the F1 function key followed by one of the arrow keys. This procedure allows you to navigate vertically 11 lines at a time (using the up- or down-arrow key) or horizontally 30 columns at a time (using the left- or right-arrow key). Practice using the F1/arrow keys combination to change the viewing area.

Notice that the report in Figure 6-1 consists of text and fields descriptions. Text and field descriptions can be entered by SELECTing the appropriate beginning location and typing the desired text or field description. Let's type the first three heading lines of the report. SELECT the appropriate beginning location and type the character string *FOR DAY ENDING MM/DD/YY*. Now use the mouse to SELECT the beginning location of the next character string and type the string *ORDER RESPONSE REPORT*. Next, SELECT the location for page and type *PAGE ZZ9*. Repeat the procedures to complete the next two header lines of the report (as illustrated in Figure 6-5).

104

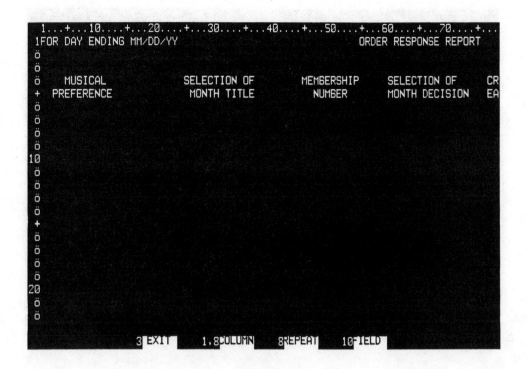

Figure 6-5

Report Design Drawing Screen

Step 4. You are now ready to enter the first detail line of the report. We could enter the field descriptions in the same manner as discussed in Step 3. However, we recommend the following approach since it exploits the power of the data dictionary.

If you have been entering data element descriptions into the diction prior to working with this report designer, we can instruct EXCELERATOR to look up the data element description in the data dictionary, thereby ensuring consistency of output pictures. Here's how. Position the cursor to the beginning location for the first field that is to appear on the detail line (*MUSICAL PREFERENCE*). SELECT the F10 **FIELD** command.[1]

EXCELERATOR now asks for the **Related Data Element Name**. Chances are you don't remember the exact name or even whether the data element was entered into the dictionary. Don't worry! Press the F4 **DESCRIBE** function key. A Selector List similar to the one illustrated in Figure 6-6 should appear (assuming that elements were previously described in the dictionary). Element Selector Lists tend to be rather lengthy. If the list is longer than one full screen, a plus sign appears at the bottom of the screen. SELECTing the plus sign displays the next full screen of elements. If you find your element on any screen, SELECT it. If the element is not found in the list, CANCEL. This takes you back to the prompt **Related Data Element Name**. Typing over the asterisk, enter the name of the new element, and press the F4 function key.

This takes you either to an existing Data Element Description or to a blank Data Element Description Screen. You initiate entries for a new element or modify, if necessary, the entries

---

1   In the report and screen design lesson, we have continued to use the mouse to SELECT commands. Alternatively, you can press the corresponding function keys.

```
Element

 Entity name                           Alternate Name

ACCEPT OFFER? . . . . . . . . . . . .
AMOUNT................................AMOUNT
CATALOG NUMBER. . . . . . . . . . .
CITY..............................
CREDIT. . . . . . . . . . . . . . .PURCHASE CREDIT
CREDITS EARNED.......................
DATE. . . . . . . . . . . . . . .
MEDIUM.............................
MEMBER NUMBER . . . . . . . . . . .MEMBER NUMBER
MUSICAL PREFERENCE...................
NAME. . . . . . . . . . . . . . . .
NO ORDER RESPONSES..................
NUMBER OF ORDER RESPONSES . . . . . .
NUMBER OF RESPONSES NOT RECEIVED.....
ORDER CREDITS . . . . . . . . . . .
ORDER DATE.........................ORDER DATE
ORDER PRICE . . . . . . . . . . . .UNIT ORDER PRICE
 . . . . . . . . . . . +  . . . . . . . . . . . . . . . .
```

Figure 6-6

Selector List for Data Elements

for an existing element. When you're finished, press the F3 function key to save and exit the dictionary description. Press F3 again and you'll see the picture clause has been copied from the dictionary to your report design. In this fashion, you maintain consistency with known data dictionary definitions of data elements.

Now use either of the above approaches to complete the field descriptions for the first detail line of the report. We should point out that we prefer not to document fields such as the report date, page number, and message in a data dictionary since they are not normally stored in files. Instead, we opt to enter those field descriptions as text (see Step 3).

Step 5.   You now know the basics required to complete the remainder of the report. However, there are some opportunities for reducing time and effort. Notice that some field descriptions should appear over several lines of the report (for example, *CATALOG NUMBER* and *MEDIA*). Rather than individually describe each additional occurrence of those fields on subsequent lines, you can choose one of two easier ways.

The first approach is used when a field is to be repeated over a common column of the report. Position the cursor at the field description appearing under the report column heading *CATALOG NUMBER*. SELECT **COLUMN** from the commands menu. You are prompted to *Enter number of times to repeat field*. Type 2 and then press the F3 function key. The field description should now appear two more times under the same column of the report. All repeated occurrences will be single spaced (you can't change the spacing).

Let's examine a second approach to repeating fields. SELECT the same field description, under the heading *CATALOG NUMBER*. SELECT **REPEAT** to indicate that the field is to be repeated. The system acknowledges the selection with the message *Repeat key now refers*

106

to current field. Now SELECT the beginning location where the field is to be repeated (copied). SELECT **REPEAT** again. The field description should have been copied to the new location. Let's try it again. SELECT another location repeating the same field description (note that EXCELERATOR remembers what field is to be repeated). SELECT **REPEAT** and the field description should have been copied once again.

This approach can also be used to repeat text portions of the report. For example, notice in Figure 6-1 that the text string *TOTAL ACCEPTING MONTHLY TITLE* appears three times. Unfortunately, EXCELERATOR cannot repeat text strings that include spaces. Therefore each word of the text string *TOTAL ACCEPTING MONTHLY TITLE* would have to be repeated separately. Alternatively, you could simply retype the text string at the new locations.

While you can repeat fields and text, report lines *cannot* be repeated. You can now complete the remaining portions of the report design.

Step 6.   Now save your work. To save your work SELECT **EXIT** from the drawing commands menu. Your report is now saved and the report design Action Keypad should reappear.

Exercise 6.2   Prototype a Report

Exercise 6.1 is great if you have a pretty good feel for the report to be designed. Suppose, instead, that management can only give you a concept for the report but not describe the specific layout. In this situation, you can use EXCELERATOR to prototype the actual report. We're going to start from scratch and prototype the report from Exercise 6.1. The prototype will look like Figure 6-7. After the report is prototyped, we can convert it to a design specification for programmers (as done in Exercise 6.1).

The prototype is intended to serve as a model or realistic representation of the report to users. Therefore, our prototype version of the report differs primarily from a report design in that it will contain sample data instead of picture clauses. To realistically demonstrate the approach, you will create your prototype from scratch, *not* using the design specification you created in Exercise 6.1.

It is important to acknowledge that the procedures for creating the report prototype are no different from those described earlier in creating the report design (except field descriptions will be replaced with data values entered as text).

Step 1.   From the Main Menu, SELECT **SCREENS & REPORTS**. From the sub-menu, SELECT **Report Design**. SELECT **Add** from the Action Keypad. For this report, type the name *ORDER RESPONSE REPORT PROTOTYPE*. Press the Enter key.

Step 2.   A blank Report Design Description Screen should appear. Add the same description used in Exercise 6.1: *THIS REPORT DETAILS THE ORDER RESPONSES RECEIVED DURING A SINGLE DAY. THE REPORT IS GROUP INDICATED TO DIFFERENTIATE ORDERS BY MEMBER AND MUSICAL PREFERENCE CATEGORY.* SELECT the **Created By:** entry. Type your name.

Again, let's use the default report size. Press the F3 function key to save, exit, and proceed to the Report Design Drawing Screen.

To draw the report in Figure 6-7 you must move the cursor to the appropriate location(s) within the drawing area. Cursor positioning was described in the previous exercise.

Notice that the report in Figure 6-7 consists of text and sample data. Text and data can be entered by SELECTing the appropriate beginning location and typing the desired text or data. Type the first three heading lines of the report.

FOR DAY ENDING 03/31/87   ORDER RESPONSE REPORT   PAGE   4

MUSICAL PREFERENCE	SELECTION OF MONTH TITLE	MEMBERSHIP NUMBER	SELECTION OF MONTH DECISION	CREDITS EARNED	CATALOG NUMBER	MEDIA	MESSAGES
COUNTRY	THE GAMBLER	30214	YES	3	33102	CASSETTE	
					45211	CASSETTE	
					65005	CASSETTE	
		39449	NO	1	33801	RECORD	
		43878	NO	1	12389	RECORD	
		54098	NO	0			NO ORDER
		66603	NONE	1	33102	RECORD	AUTOMATIC ORDER
		76075	YES	1	33102	8TRACK	
		77443	NO	0			NO ORDER
		80937	YES	7	33102	RECORD	
					34044	COMPACT DISC	
					60438	RECORD	
					79870	RECORD	
					90994	COMPACT DISC	
		90043	YES	2	33102	RECORD	
					50511	RECORD	

TOTAL ACCEPTING MONTHLY TITLE   4
TOTAL REJECTING MONTHLY TITLE   4

EASY LISTENING	WHAT'S NEW	44082	YES	1	43366	RECORD	
		60118	NO	0			NO ORDER
		74844	NO	6	20112	COMPACT DISC	
					54430	COMPACT DISC	
					98303	COMPACT DISC	
		75593	YES	1	43366	RECORD	
		80326	YES	2	43366	RECORD	
					66043	RECORD	
		87620	NONE	1	43366	RECORD	AUTOMATIC ORDER
		90093	YES	1	43366	CASSETTE	
		93202	NONE	1	43366	RECORD	AUTOMATIC ORDER

TOTAL ACCEPTING MONTHLY TITLE   4
TOTAL REJECTING MONTHLY TITLE   2

NUMBER AND PERCENT OF RESPONSES FOR ORDER   45   56%
NUMBER AND PERCENT OF RESPONSES FOR NO ORDER   36   44%
NUMBER OF AUTOMATIC ORDERS (NONRESPONDENTS)   12

TOTAL NUMBER OF ORDERS   81

Figure 6-7

Order Response Report Prototype

108

Step 3. You are now ready to enter the first detail line of the report. Although you could enter the sample data exactly as demonstrated for text and headings, it would be nice if the data was consistent with output picture clauses and value ranges. That would add realism to the prototype. If you have been entering data element descriptions into the dictionary prior to working with this report designer, you can use the power of the dictionary to create realistic sample data. Here's how!

Position the cursor to the beginning location for the first field that is to appear on the detail line. SELECT the FIELD command. EXCELERATOR asks for the **Related Data Element Name**. Press the F4 **Describe** function key. A Selector List should appear (assuming that elements were previously described in the dictionary). If you find your element in the list, SELECT it. Press the F4 function key to call up the dictionary description for that element. Note the edit rules. Press the F3 function key to exit the dictionary description. Press F3 again and you'll see the picture clause has been copied from the dictionary to your report. With the edit rules in mind, type a valid data sample over the field description. In this fashion, you maintain consistency with known data dictionary definitions of data elements.

What if you didn't find your element in the Selector List? CANCEL to exit the selector list. You now have two options. First, you could type any data value you deem to be reasonable. Second, you could initiate a new dictionary description using the technique you learned in Exercise 6.1 and then type valid data over the resulting field description.

Complete the prototype with the above techniques. Then SELECT **EXIT** to leave the drawing screen. Your prototype has been saved and you have returned to the Action Keypad for Report Design.

The prototype can now be reviewed with the users. The users will likely suggest improvements to which you would respond by modifying the prototype. Once the prototype has been approved, you could convert it to a design as using the technique you learned in Exercise 6.1.

Step 4. Now save your work. To save your work SELECT **EXIT** from the drawing commands menu. Your report is now saved and the report design Action Keypad should appear. SELECT **Exit** to return to the Main Menu.

Step 5. If you don't plan to proceed immediately with the next lesson, you might consider doing a **Backup** (via **HOUSEKEEPING**). Printing and backup were covered in Lesson 3.

Exercise 6.3 Prototype Screens

EXCELERATOR significantly aids in the rapid prototyping of input, output, and dialogue screens. In this exercise you will learn how to use EXCELERATOR to define screens, link multiple screens, and test or exercise the screens. We will recreate the screens appearing in Figures 6-8 and 6-9. The screen appearing in Figure 6-8 is a menu/dialogue screen used by order entry to perform a variety of options associated with processing orders. The screen in Figure 6-9 represents an input screen used to enter data about a new order. Let's begin by creating the screen in Figure 6-8.

Step 1. If necessary, log on to EXCELERATOR and **Restore** (via **HOUSEKEEPING**) the work you've completed in Lessons 3 through 6. SELECT **SCREENS & REPORTS** from the Main Menu. SELECT **Screen Design** from the sub-menu that appears to the right.

Step 2. A screen design Action Keypad appears. SELECT **Add**. Let's call our screen *ORDER MENU*. Press the Enter key.

Step 3. A Screen Design Description Screen should appear (Figure 6-10). The cursor is positioned at **Next Screen Name**. Type *ORDER* in preparation for the second screen that you'll design.

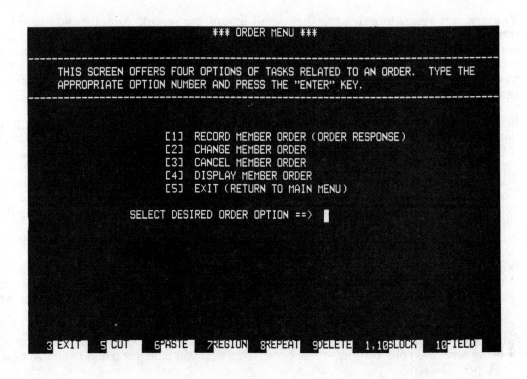

Figure 6-8

Order Menu Prototype

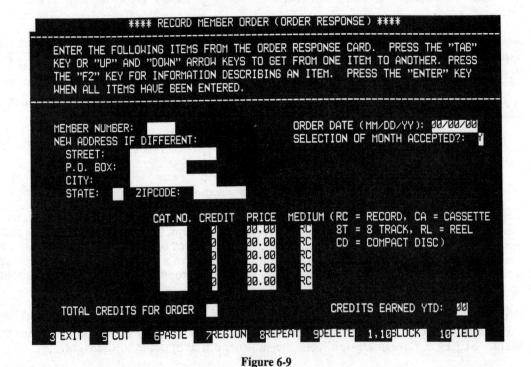

Figure 6-9

Order Entry Screen Prototype

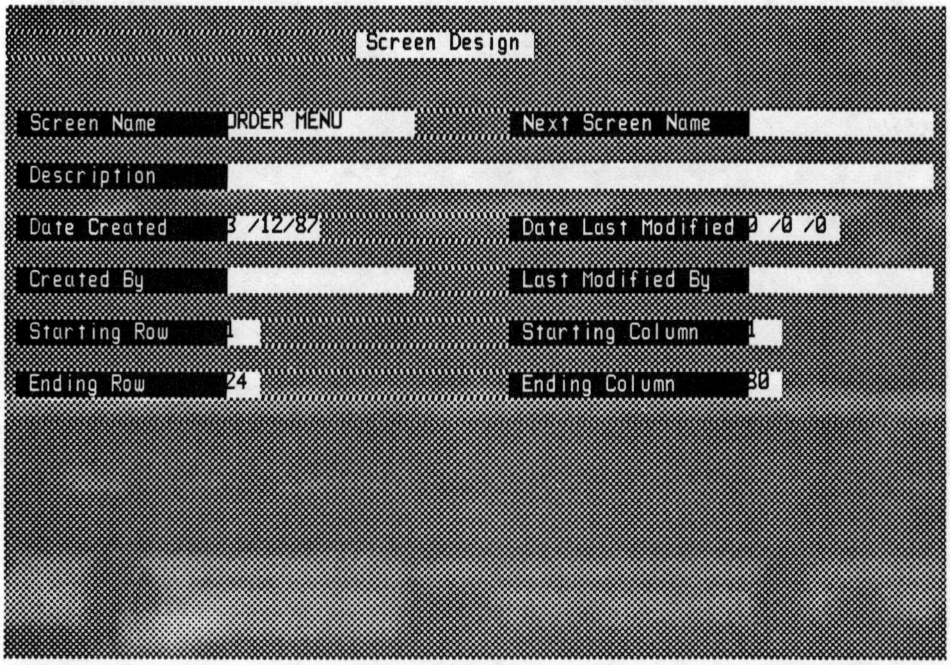

Figure 6-10

Screen Design Description Screen

Press the Enter key to get to the **Description**. Type *THIS SCREEN PROVIDES ACCESS TO SEVERAL ORDER PROCESSING OPTS*. SELECT the **Created By** field and type your name. We will use the current screen size default values for this exercise. Press the F3 function key to save your work.

Step 4. You should now see a blank Screen Design Drawing screen. At the bottom of the drawing screen is a command menu. Entering text (e.g., title, instructions, menu option numbers, menu option descriptions, etc.) for a screen is accomplished in a similar manner to entering text for report designs (covered in Exercise 6.1). Type the text portion of the screen in Figure 6-8. Simply position the cursor to the appropriate location(s) and begin typing.

Step 5. Your menu screen is to contain a single field for the user to enter the desired option. Let's add that field to the screen. SELECT the position for that field. Now SELECT FIELD from the command menu. A **Field Definition Screen** should appear at the bottom of the drawing screen (Figure 6-11). This screen is used to describe the characteristics of our field.[2] Instructions and explanations for completing the screen are as follows:

a. The cursor is at **Field Name**. Type *ORDER OPTION*. Press the Enter key to get to **Related ELE**.
b. **Related ELE** is used to establish a link to a data element in the dictionary. Since the field we are describing doesn't exist in the dictionary, press the Enter key to go to **Length**.
c. **Length** is used to indicate the number of spaces the field will occupy on the screen (including decimals, dollar signs, etc.). For our field, type the number *1*. Press the Enter key.

---

2   To familiarize yourself with the entire screen, you can press the Tab key to move from attribute-to-attribute, stopping at selected attributes to press the F2 function key for help.

111

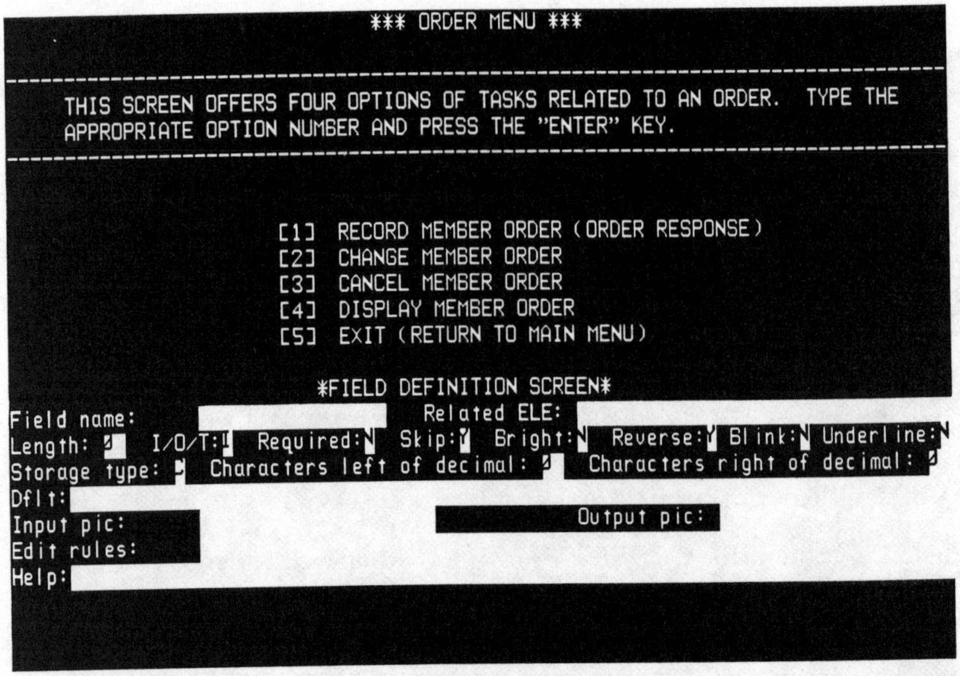

```
                    *** ORDER MENU ***
-----------------------------------------------------------
    THIS SCREEN OFFERS FOUR OPTIONS OF TASKS RELATED TO AN ORDER.  TYPE THE
    APPROPRIATE OPTION NUMBER AND PRESS THE "ENTER" KEY.
-----------------------------------------------------------



            [1]  RECORD MEMBER ORDER (ORDER RESPONSE)
            [2]  CHANGE MEMBER ORDER
            [3]  CANCEL MEMBER ORDER
            [4]  DISPLAY MEMBER ORDER
            [5]  EXIT (RETURN TO MAIN MENU)

                  *FIELD DEFINITION SCREEN*
Field name:                    Related ELE:
Length:     I/O/T:   Required:N  Skip:Y  Bright:   Reverse:Y Blink:N Underline:N
Storage type:    Characters left of decimal: 9   Characters right of decimal: 9
Dflt:
Input pic:                              Output pic:
Edit rules:
Help:
```

Figure 6-11

Field Definition Window in the Screen Drawing Area

d. The **I/O/T** entry indicates whether the field will be used to input or enter data values (I), output system-provided data or variable messages (O), or as a text character string for fixed messages that cannot be changed (T). The *ORDER OPTION* field will be used to input the user option. Press *I* to set the value. The cursor automatically moves to the next position.

e. **Required** is a *yes/no* field. We require the user to enter a value; therefore, type *Y.* The cursor automatically moves to the next position.

f. The **Skip** attribute is used to indicate whether or not the cursor should automatically skip to the next field once the user has typed data values in the current field. (note: all positions in the current field must have been filled.) When our user enters any value, we want to proceed immediately; therefore, type *Y.* The cursor will automatically move to the next entry.

g. The next four locations represent display attributes for the field. They control bright (boldface), reverse video, blinking field, and underlined display. The values for all four entries are *Y* or *N* and the cursor automatically moves to the next entry. The current defaults are acceptable for our screen.

h. **Storage type** indicates how a value entered for the field will be stored. The permissible values are *B* (binary), *C* (character), *D* (date), *F* (floating point), and *P* (packed decimal). Type *C* for our menu.

i. **Characters to the left of decimal** indicates the display length of character and integer fields. Type *1* and press the Enter key.

j. **Characters to the right of decimal** is used only for real numbers (**Storage Type** = *F* or *P*). This is not applicable to our field; therefore, press the Enter key.

k. **Dflt** stands for *default value* for the field. It will be displayed in the input field. We don't want a default; therefore, press the Enter key.

l. **Input pic:** is used to describe the input editing format for the field. EXCELERATOR uses a COBOL-like syntax for specifying input editing formats. The edit formats were

112

covered in Lesson 5, Exercise 5.3. Our field represents a one digit alphanumeric; therefore, type the number *9*. Press the Enter key.

m.   **Output pic:** is used to describe the format or appearance of an output field. It is similar to that of the input picture. Our field is not an output field; therefore, there appears to be no need to provide an entry for this attribute. However, EXCELERATOR will eventually generate a COBOL Data Division for our screen and it uses the Output pic to do that. Therefore, we recommend that you specify an output picture identical to the input picture. Type *9* and press the Enter key.

o.   **Edit rules:** defines the editing requirements to be applied to the field. The edit rules are extensive and are covered in Lesson 5, Exercise 5.3. When a user enters a value for *ORDER OPTION*, we want to accept the numeric values 1 through 5. Type *VALUES ARE 1,2,3,4,5.* Then press the Enter key.

p.   **Help** defines a message or information describing the current field. It can be displayed by the user when the cursor is positioned at the field by pressing the F2 key. Provide the user with a brief explanation of the *ORDER OPTION* field. Type *ONE OF 5 OPTIONS FROM THE ORDER MENU. VALUE IS A 1 DIGIT NUMBER.* Since we are finished with our field description, press the F3 function key.

The field description for *ORDER OPTION* should now appear on the drawing screen as a single-character, highlighted block. You'll see this if you move the cursor elsewhere on the screen.

Step 6.   Now save your work. To save your work SELECT **EXIT** from the drawing commands menu. Your screen is now saved and the screen design Action Keypad should appear.

You're now ready to complete the input screen in Figure 6-9. Recall that this screen is linked with the *ORDER MENU* screen you just created. This screen will be viewed if the user selects Option 1 from the menu. It represents an input screen used to capture new orders received in the form of an order response card. Let's recreate the screen.

Step 7.   SELECT **ADD** from the screen design Action Keypad. You are now prompted to enter the name of the new screen. Type *ORDER* and then press the Enter key.

Step 8.   A blank Screen Design Description Screen should appear. You already know how to respond to this screen. There is no **Next Screen Name** for this exercise. Press the F3 key. You should now see a blank Screen Design Drawing Screen.

Step 9.   Enter the screen header and instruction lines (lines 1 through 7) at the top of the screen. Those lines should be entered as text. You already know how to do this.

Now you are ready to define fields to be included on the screen. For your convenience, the field definitions are provided in Figure 6-12.

Step 10.   Position the cursor at the appropriate location on line 9 and type the text *MEMBER NUMBER*. Now space over two positions and let's define a field for the user to input data values for *MEMBER NUMBER*. It is important to note that the first field (upper left-most) appearing on the screen should be the key field

SELECT **FIELD** and a field definition screen should appear. This screen should look familiar since you completed a similar screen earlier in this exercise. However, if you did Lesson 5 (*Information Engineering*), then a data dictionary entry for *MEMBER NUMBER* already exists. Therefore, the Field Definition Screen can be used to pull in the values defined in Lesson 5. Here's how.

a.   The first thing to do is to find out if *MEMBER NUMBER* already exists in the dictionary. For **Field Name** type *MEMBER NUMBER*. Press the Enter key. For Related ELEment, check for an existing data element description. Press the F4 function key. A Selector List of elements appears.

Field groups

Name:ORDER ITEM   Start row:17 End row:21 Total rows:10
                    Start col:23 End col:48

Field Descriptions

Input Fields

Row,col: 9, 21 Field name: MEMBER NUMBER
Len: 5 I/O/T:I Req:Y Skip:Y Brt:N Rev:Y Blnk:N Und:N Stor:C(5, 0)
Dflt:
Input pic:99999   Output pic:99999
Edit rules:
Help:A 5 DIGIT NUMBER THAT UNIQUELY IDENTIFIES A CLUB MEMBER.

Row,col: 9, 69 Field name: ORDER DATE
Len: 8 I/O/T:I Req:N Skip:Y Brt:N Rev:Y Blnk:N Und:N Stor:D(8, 0)
Dflt:00/00/00
Input pic:MMDDYY   Output pic:MM/DD/YY
Edit rules:
Help:THE DATE FROM ORDER RESPONSE CARD IN MM/DD/YY FORMAT.

Row,col: 10, 77 Field name: ACCEPT OFFER?
Len: 1 I/O/T:I Req:Y Skip:Y Brt:N Rev:Y Blnk:N Und:N Stor:C(1, 0)
Dflt:Y
Input pic:X   Output pic:X
Edit rules:VALUES ARE "Y","N"
Help:INDICATES WHETHER INDIV. ACCEPTS THE MONTHLY SELECTION OFFER

Row,col: 11, 18 Field name: STREET
Len: 15 I/O/T:I Req:N Skip:Y Brt:N Rev:Y Blnk:N Und:N Stor:C(15, 0)
Dflt:
Input pic:X(15)   Output pic:X(15)
Edit rules:
Help:STREET OF MEMBER'S RESIDENCE.

Row,col: 12, 18 Field name: P.O. BOX
Len: 10 I/O/T:I Req:N Skip:Y Brt:N Rev:Y Blnk:N Und:N Stor:C(10, 0)
Dflt:
Input pic:X(10)   Output pic:X(10)
Edit rules:OPTIONAL
Help:P.O. BOX FOR MAILING TO MEMBER'S RESIDENCE

Row,col: 13, 18 Field name: CITY
Len: 15 I/O/T:I Req:N Skip:Y Brt:N Rev:Y Blnk:N Und:N Stor:C(15, 0)
Dflt:
Input pic:X(15)   Output pic:X(15)
Edit rules:
Help:CITY OF MEMBER'S RESIDENCE

Row,col: 14, 15 Field name: STATE
Len: 2 I/O/T:I Req:N Skip:Y Brt:N Rev:Y Blnk:N Und:N Stor:C(2, 0)
Dflt:
Input pic:AA   Output pic:AA
Edit rules:FROM "STATE CODE"
Help:A TWO CHARACTER CODE OF U.S. STATE (SEE STATE CODE TABLE)

Row,col: 14, 29 Field name: ZIPCODE
Len: 9 I/O/T:I Req:N Skip:Y Brt:N Rev:Y Blnk:N Und:N Stor:C(9, 0)
Dflt:
Input pic:X(9)   Output pic:X(9)
Edit rules:
Help:A 5 TO 9 CHARACTER IDENTIFYING POSTAL AREA OF RESIDENCE.

Row,col: 17, 23 Field name: CATALOG NUMBER
Len: 5 I/O/T:I Req:N Skip:Y Brt:N Rev:Y Blnk:N Und:N Stor:C(5, 0)
Dflt:
Input pic:99999   Output pic:99999
Edit rules:
Help:A 5 DIGIT NUMBER UNIQUELY IDENTIFYING AN INVENTORIED ITEM.

Row,col: 17, 32 Field name: CREDIT
Len: 1 I/O/T:I Req:N Skip:Y Brt:N Rev:Y Blnk:N Und:N Stor:C(1, 0)
Dflt:0
Input pic:9   Output pic:9
Edit rules:
Help:NUMBER OF CREDITS EARNED FOR PURCHASE OF ITEM.

Row,col: 17, 38 Field name: ORDER PRICE
Len: 5 I/O/T:I Req:N Skip:Y Brt:N Rev:Y Blnk:N Und:N Stor:C(5, 0)
Dflt:00.00
Input pic:99.99   Output pic:99.99
Edit rules:VALUES ARE 0 THRU 100
Help:THE PRICE OF A PRODUCT AT TIME OF ORDER.

Row,col: 17, 47 Field name: MEDIUM
Len: 2 I/O/T:I Req:N Skip:Y Brt:N Rev:Y Blnk:N Und:N Stor:C(2, 0)
Dflt:RC
Input pic:XX   Output pic:XX
Edit rules:VALUES "RC","CD","8T","RL","CA"
Help:AVAILABLE MEDIUM FOR ITEM (E.G. RECORD, CASSETTE, REEL...) ....)

**** Remaining 4 lines of this scroll region are skipped *****

Figure 6-12(a)

Field Definitions for Screens (Part 1)

Text:THE "F2" KEY FOR INFORMATION DESCRIBING AN ITEM. PRESS THE "ENTER" KEY

Row,col: 6, 5
Len: 33 I/O/T:T Req:N Skip:N Brt:N Rev:N Blnk:N Und:N
Text:WHEN ALL ITEMS HAVE BEEN ENTERED.

Row,col: 7, 1
Len: 80 I/O/T:T Req:N Skip:N Brt:N Rev:N Blnk:N Und:N
Text:——

Row,col: 9, 5
Len: 14 I/O/T:T Req:N Skip:N Brt:N Rev:N Blnk:N Und:N
Text:MEMBER NUMBER:

Row,col: 9, 46
Len: 22 I/O/T:T Req:N Skip:N Brt:N Rev:N Blnk:N Und:N
Text:ORDER DATE (MM/DD/YY):

Row,col: 10, 5
Len: 70 I/O/T:T Req:N Skip:N Brt:N Rev:N Blnk:N Und:N
Text:NEW ADDRESS IF DIFFERENT: SELECTION OF MONTH ACCEPTED?:

Row,col: 11, 7
Len: 7 I/O/T:T Req:N Skip:N Brt:N Rev:N Blnk:N Und:N
Text:STREET:

Row,col: 12, 7
Len: 9 I/O/T:T Req:N Skip:N Brt:N Rev:N Blnk:N Und:N
Text:P.O. BOX:

Row,col: 13, 7
Len: 5 I/O/T:T Req:N Skip:N Brt:N Rev:N Blnk:N Und:N
Text:CITY:

Row,col: 14, 7
Len: 6 I/O/T:T Req:N Skip:N Brt:N Rev:N Blnk:N Und:N
Text:STATE:

Row,col: 14, 19
Len: 8 I/O/T:T Req:N Skip:N Brt:N Rev:N Blnk:N Und:N
Text:ZIPCODE:

Row,col: 16, 22
Len: 57 I/O/T:T Req:N Skip:N Brt:N Rev:N Blnk:N Und:N
Text:CAT.NO. CREDIT PRICE MEDIUM (RC = RECORD, CA = CASSETTE

Row,col: 17, 53
Len: 23 I/O/T:T Req:N Skip:N Brt:N Rev:N Blnk:N Und:N
Text:8T = 8 TRACK, RL = REEL

Row,col: 18, 53
Len: 18 I/O/T:T Req:N Skip:N Brt:N Rev:N Blnk:N Und:N
Text:CD = COMPACT DISC)

Row,col: 23, 0
Len: 23 I/O/T:T Req:N Skip:N Brt:N Rev:N Blnk:N Und:N
Text:TOTAL CREDITS FOR ORDER

Row,col: 23, 52
Len: 19 I/O/T:T Req:N Skip:N Brt:N Rev:N Blnk:N Und:N
Text:CREDITS EARNED YTD:

Output Fields

Row,col: 23, 31 Field name: ORDER CREDITS
Len: 2 I/O/T:O Req:N Skip:Y Brt:N Rev:Y Blnk:N Und:N Stor:C(2, 0)
flt:
Output pic:Z9

Row,col: 23, 73 Field name: CREDITS EARNED
Len: 2 I/O/T:O Req:Y Skip:Y Brt:N Rev:Y Blnk:N Und:N Stor:C(2, 0)
flt:00
Output pic:Z9

Text Fields

Row,col: 1, 18
Len: 46 I/O/T:T Req:N Skip:N Brt:N Rev:N Blnk:N Und:N
Text:•••• RECORD MEMBER ORDER (ORDER RESPONSE) ••••

Row,col: 2, 1
Len: 80 I/O/T:T Req:N Skip:N Brt:N Rev:N Blnk:N Und:N
Text:——

Row,col: 3, 5
Len: 72 I/O/T:T Req:N Skip:N Brt:N Rev:N Blnk:N Und:N
Text:ENTER THE FOLLOWING ITEMS FROM THE ORDER RESPONSE CARD. PRESS THE "TAB"

Row,col: 4, 5
Len: 72 I/O/T:T Req:N Skip:N Brt:N Rev:N Blnk:N Und:N
Text:KEY OR "UP" AND "DOWN" ARROW KEYS TO GET FROM ONE ITEM TO ANOTHER. PRESS

Row,col: 5, 5
Len: 71 I/O/T:T Req:N Skip:N Brt:N Rev:N Blnk:N Und:N

Figure 6-12(b)

Field Definitions for Screens (Part 2)

115

b.   If a related data element appears in the list (even under a different name), SELECT that element. In most cases, you would not alter the data dictionary's definition, with the possible exception of customizing the pictures and help message. You might need to increase length to accommodate edit masks such as adding periods, slashes, dollar signs, etc. Finally, you would need to determine the values for **I/O/T**, **Required**, **Skip**, **Bright**, **Reverse**, **Blink**, and **Underline**, since they are not recorded in the data element dictionary. Press the F10 function key. This transfers a description from the dictionary element to the field. You can edit the field definition screen without affecting the data element dictionary. Finally, press F3 to establish the field on the prototype screen.

c.   If you do not find a related data element in the Selector List, and you wish to enter the field into the dictionary as an element, CANCEL the selector list. In **Related ELE**, type the name of the new element (usually the same name as the field). Press F4, the **Describe** function key. This takes you to the data dictionary screen. Complete the description (covered in Lesson 5). Press F3 to save the dictionary description. To complete the field definition form from the new dictionary entry, press the F10 function key. The field definition is complete. Finally, press F3 to establish the field on the prototype screen.

Step 11.   Now complete the remaining text and fields appearing *on lines 9 through 16*. Don't complete the line containing the catalog number, credit, price, and medium fields.

Step 12.   Notice that line 16 of our screen contains four fields that repeat five times. In reality these fields repeat ten times. These fields are set up such that only five occurrences of the fields display at any one time. The additional five occurrences can be seen by the user by scrolling. This is commonly referred to as a *scrolling region* on the screen. Thus, the user can input up to ten items from an order response card. Here's how to set up this feature of the screen.

First, define *each* of the four fields to appear on line 17 (under the column headings). For ease of explanation, use the precise Row and Column locations specified for each field in Figures 6-12. SELECT the left-most position of the first field (column 23, row 17). SELECT **REGION**. A Region Definition Screen should appear (Figure 6-13).

The first position asks for a Region name. Type ORDER ITEM and press the Enter key to go to the next location, Lines displayed. Type 10 and press the Enter key. You are now at Total columns. The total number of columns required by our scroll region is 26 (the width of the scroll region). Type 26. That completes the Region Definition Screen; therefore, press the F3 function key. The scroll region should appear on your prototype.

Step 13.   Now complete the remaining portions of the screen (the total line).

Step 14.   Now save your work. To save your work SELECT **EXIT** from the drawing commands menu. Your screen is now saved and the screen design Action Keypad should reappear. If you don't plan to proceed immediately to Exercise 6.4, SELECT **Exit** from the Screen Design Menu and then SELECT **Exit** from the Screens & Reports Menu. Also, if you don't plan to proceed immediately to the next lesson, you might consider doing a **Backup** (via **HOUSEKEEPING**). Printing and backup were covered in Lesson 3.

Exercise 6.4   Test Screen Prototypes

If you've covered Exercise 6.3, you are now ready to test your screens. EXCELERATOR's **Screen Design** facility provides the ability to test (or **Inspect**) a screen or series of linked screens. While you are learning how to test your screens, keep in mind that the approach is appropriate for directly involving users in the testing process.

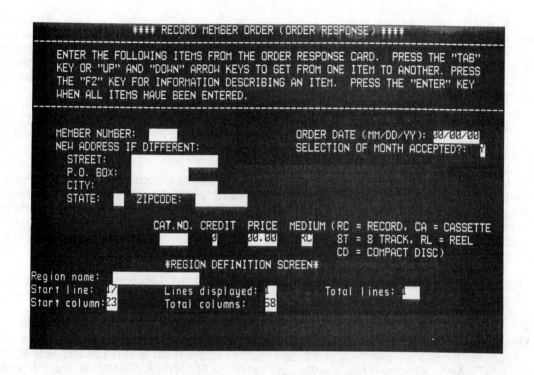

Figure 6-13

Region Definition Window In the Screen Design Area

Step 1. If necessary, log on to EXCELERATOR and **Restore** (via **HOUSEKEEPING**) the work you've completed thus far. SELECT **SCREENS & REPORTS** from the Main Menu. SELECT **Screen Design** from the sub-menu that appears to the right.

Step 2. A screen design Action Keypad should appear. We want to test both screens created in Exercise 6.3. Select **Inspect**. You are now asked to enter the name of a screen design. Since we want to test both screens created in Exercise 6.3, and since those two screens are linked, we'll provide the name of the first screen. Type *ORDER MENU* and press the Enter key.

Step 3. The *ORDER MENU* screen should appear. There's only one field to test on this screen. Let's test it. First, let's type an invalid option number. Type the number 8. If things work correctly, you get an error message indicating that the option number was invalid. Now let's suppose the user was still confuse. He or she should be able to get additional help or instructions concerning the field. Press the F2 function key. The message for the Help attribute of the Field Definition Screen should appear. Now let's see if the screen accepts a valid option for the field. Type the number 1 (or any number from 1 to 5). The *ORDER* screen should now appear.

Step 4. Now let's test the *ORDER* screen. If you used Figure 6-12 to complete your field definitions, you may also want to refer to them to generate valid and invalid test data.

One final word concerning your test of the ORDER screen. Don't be surprised that the cursor is never positioned such that the two fields on the last line can be tested. Those fields are defined as output fields. Thus, the user should not be permitted to input data values for them.

117

Exercise 6.5   Exercise Screens and Create Test Data

One of the most time-consuming, difficult, and neglected tasks for a systems analyst is the generation of test data. With EXCELERATOR the process is made much easier. Within EXCELERATOR, you can create test data while actually testing screens. In this exercise you'll learn how to create test data for the *ORDER* screen created in Exercise 6.3. More important, at the same time you will be learning an alternative to Exercise 6.4 for testing screens.

Step 1.   If necessary, log on to EXCELERATOR and **Restore** (via **HOUSEKEEPING**) the work you've completed thus far. SELECT **SCREENS & REPORTS** from the Main Menu. SELECT **Screen Data Entry** from the sub-menu that appears to the right.

Step 2.   A screen data entry Action Keypad appears. We want to exercise the *ORDER* screen created in Exercise 6.3. In addition, we want to create sample test data for the *ORDER* screen. To do so, you must first define a file that will be used to store data values entered through the screen. SELECT **Add** from the Action Keypad. You are now asked to enter the name of the data file. Type *SAMPLE ORDER* and press the Enter key. A new Screen Data Entry Screen should appear (Figure 6-14).

Step 3.   Type ORDER. Press the Enter key. EXCELERATOR will calculate the Record Length. Press the Enter key. Type 5, the key length of the *MEMBER NUMBER* field. Press the F3 function key to save and exit.

Step 4.   From the Action Keypad SELECT **Execute**. At the **Data File** prompt, type *SAMPLE ORDER*. Press F3. You now see a new Action Keypad for **SAMPLE ORDER RECORDS**. SELECT **Add**. After entering a valid data value for *MEMBER NUMBER*, the *ORDER* screen should appear immediately. You can now test this screen in the same manner as you did in Exercise 6.4. When your testing is complete, press the F3 key.

Step 5.   The *SAMPLE ORDERS RECORDS* Action Keypad should reappear. Testing of the *ORDER* screen is complete, and a sample test data record has been saved in the *SAMPLE ORDER* file. You may wish to create additional test data records using the *ORDER* screen. To do so, repeat Step 4 for each test data record you wish to create.

Step 6.   If you wish, you can get a screen or printer output of your screen and test data. SELECT **Output** from the Action Keypad. You're now asked to enter a value for the **record key**. We can enter a record key value and get an output of our screen containing that particular record's data values. However, let's get an output of your screen for all records contained in your *SAMPLE ORDER* file. Type an * (asterisk) and press the F3 key. You are now asked where the output is *to go*. SELECT **Screen**. *Your screen with sample test data from the SAMPLE ORDER file* should now appear. Press any key to return to the Action Keypad. Now SELECT **Exit** and then SELECT **Exit** again to return to the **SCREENS & REPORTS** Menu.

Step 7.   If you don't plan to proceed immediately to the next lesson SELECT **Exit** to return to the Main Menu. Also, you might consider doing a **Backup** (via **HOUSEKEEPING**). Printing and backup were covered in Lesson 3.

Exercise 6.6   Generate Code for Record Layouts

In this exercise you will learn how to use EXCELERATOR to generate program code describing a record layout for screen designs. EXCELERATOR supports code generation for BASIC, C, COBOL, and PL/I programming languages.

In this exercise you will generate COBOL code for the record layout for the screens you created in Exercise 6.3. The sample COBOL code you will create is depicted in Figure 6-15.

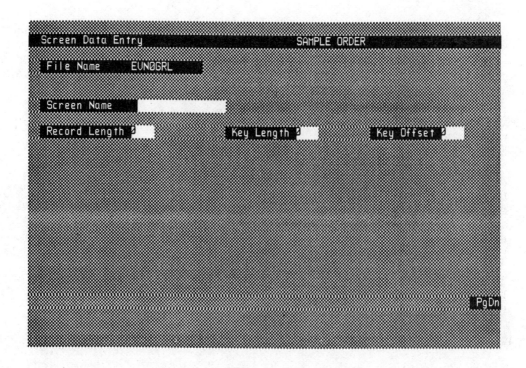

Figure 6-14

Screen Data Entry Description Screen

Step 1. If necessary, log on to EXCELERATOR and Restore (via HOUSEKEEPING) the work you've completed thus far. SELECT **SCREENS & REPORTS** from the Main Menu. SELECT **Screen Design** from the sub-menu that appears to the right.

Step 2. A screen design Action Keypad should appear. SELECT **Generate** from the Action Keypad. You are now asked to enter the name of the screen for which you want to generate program code. Type *ORDER MENU* and press the F3 function key.

Step 3. You are now asked whether you want a data map or interface file. SELECT **Data map**.

Step 4. A Generate Record Layout Screen should appear. The first location defaults to the name of your screen. You should realize that EXCELERATOR will use this name within the code describing the record layout of your screen. If you desire, you may enter a different name. Tab to the next location. This position is optional. When EXCELERATOR generates a record layout description, it will default to using the exact field or data element name you entered in defining field description for your screens. Since these same field names may appear in many other programs, you might wish to make them unique. You can do so by entering a prefix to be appended to the names. If you desire, enter a prefix. Tab to the final location on the screen. As mentioned earlier, EXCELERATOR supports code generation for several programming languages. We want a COBOL record layout description of our screens. Type *COBOL* and press the F3 function key.

119

```
*Record ORDER-MENU Compiled: 12-MAR-87
01  ORDER-MENU.
        05  ORDER-OPTION       PIC 9.
        05  MEMBER-NUMBER      PIC 99999.
        05  ORDER-DATE         ????.
        05  ACCEPT-OFFER?      PIC X.
        05  STREET             PIC X(15).
        05  P.O.-BOX           PIC X(10).
        05  CITY               PIC X(15).
        05  STATE              PIC AA.
        05  ZIPCODE            PIC X(9).
        05  ORDER-ITEM         OCCURS 10 TIMES.
            10  CATALOG-NUMBER
                               PIC 99999.
            10  CREDIT         PIC 9.
            10  ORDER-PRICE    PIC 99.99.
            10  MEDIUM         PIC XX.
        05  ORDER-CREDITS      PIC Z9.
        05  CREDITS-EARNED     PIC Z9.
```

Figure 6-15

COBOL Code Generated From Screen Design Records

Step 5.   You are now asked where you would like your output to go. Let's have the output displayed to the screen. SELECT **Screen**. The COBOL description of a record layout for the *ORDER MENU* screen should appear (Figure 6-15). Note that the record layout actually describes both the *ORDER MENU* and *ORDER* screens. That's because the two screens were linked in Exercise 6.3.

Step 6.   Now press any key, and the screen design Action Keypad reappears. For practice, repeat Steps 2 through 5 generating code for BASIC, C, and PL/I.

Step 7.   SELECT **Exit** to return to the **SCREENS & REPORTS** sub-menu. SELECT **Exit** to return to the Main Menu. If you don't plan to proceed immediately to the next lesson, you might consider doing a **Backup** (via **HOUSEKEEPING**). Printing and backup were covered in Lesson 3.

Conclusion

In this lesson, you learned how to prototype reports, screens, and simple screen dialogues using EXCELERATOR. The **SCREENS & REPORTS** facility can be very useful for demonstrating systems to users without the overhead of conventional programming.

This lesson also concludes Part Two of the tutorial. Consequently, you should be able to use EXCELERATOR to create an entire structured specification for a system. But how do you maintain the structured specification? How do you account for the complex interrelationships between specifications? How do you check the quality of specifications for consistency and completeness? These and other questions will be addressed in Part Three, *Putting the Potential to Work*.

120

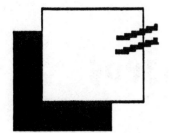

Part Three: Putting the Potential to Work for You

In Part Two, you learned how to use EXCELERATOR's facilities to create system specifications. Since that task consumes much of the traditional analyst's time, this capability represents much of the value of the product. However, the full potential of EXCELERATOR lies in some of the less understood or visible facilities. These facilities build on the general documentation facilities to maintain documentation, ensure its quality, and enhance its presentation to different audiences. They turn EXCELERATOR from a very useful package into a remarkable package. These lessons, named after the 1986 EXCELERATOR Users Conference theme, are intended to help you exploit some of the more powerful facilities.

Part Three consists of four lessons. Lesson 7 teaches you how to use the power of the data dictionary to enhance and maintain the systems specifications. Lesson 8 introduces EXCELERATOR's quality checking facility, called ANALYSIS. ANALYSIS can help you identify errors, inconsistencies, and incompleteness in your system specifications, saving you considerable time and frustration when you get to the implementation phases (where such problems tend to prove more costly to correct). Lesson 9 will teach you how to package specifications, including those from non-EXCELERATOR software packages, into easy-to-print documents for a variety of audiences. Lesson 10 concludes this tutorial with a survey of other EXCELERATOR facilities, capabilities, and value added products. This survey is intended to direct your continued education in EXCELERATOR and CASE technology.

These lessons work from the project dictionary that you created in Part Two. You should restore your dictionary before working on the lessons. As usual, you may terminate a session at any time; however, we recommend that you backup your data to prepare for the possible loss of your work.

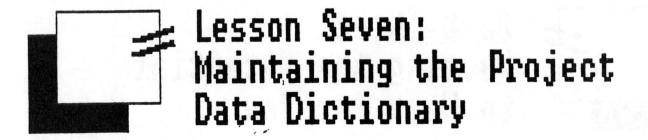

Lesson Seven:
Maintaining the Project
Data Dictionary

The Demonstration Scenario

You have developed first draft system specifications for the record club project. You were able to develop most of your specifications as an outgrowth of various system models (for instance, data flow diagrams and data model diagrams). Given these system models, you even prototyped a few screens and reports - just to give the users and management some feel for how the new system might work.

The past week has been spent in intense group walkthroughs of your specifications. Several people have been involved in the walkthroughs: users, user management, data processing management, other systems analysts, a few programmers, and the like. As is the case with most specification documents, errors and omissions were uncovered. Improvements were suggested. Naming issues were addressed. In short, you have a lot of corrections to make to your project dictionary.

Maintaining specification documents is wrought with opportunities for inconsistency. If you change the name of one data element, how many data flow, data store, and data entity records must also be changed? If you change the size or format of a data element, won't existing screen and report designs be altered? Somebody wants to rename an output (data flow). Won't that change several of your DFDs? How do you maintain all these specifications and carefully ensure that the final specification document will be consistent?

What Will You Learn in this Lesson?

Maintenance of system documentation has always been one of the most sought after but elusive goals. Everybody agrees that it is desirable to maintain systems documentation, both during and after systems development. Unfortunately, the complexity and interrelationships between systems specifications make the maintenance task both time consuming and difficult, with relatively low reliability. A single change can have an enormous ripple effect throughout the entire specification document.

In this lesson, you will learn how to use EXCELERATOR to maintain a project data dictionary, called XLDICTIONARY. You've already interacted with XLDICTIONARY in earlier lessons; however, your exposure was purposefully limited to the initial capture of specifications and simple editing. You'll also learn how to output portions of your dictionary and protect your dictionary.

EXCELERATOR also includes powerful tools for extracting and analyzing the contents of the dictionary; however, those capabilities will be deferred until Lesson 8. This lesson will focus on dictionary maintenance. After completing this lesson, you will be able to:

1.   Describe the structure and capabilities of the XLDICTIONARY in terms of the following:

   *   entities
   *   entity types

* relationships
* relationship classes
* relationship types

2. Look up entity and relationship entries in the dictionary.
3. Modify, add, and delete entities in the dictionary.
4. Track the effects of changes made to the data dictionary.
5. Prevent unauthorized changes to the dictionary.
6. Output the contents of the data dictionary.

If so desired, you will also learn how naming conventions and standards can further enhance the power of the data dictionary.

Exercise 7.1   Understand the Role and Capabilities of XLDICTIONARY

Those of you who have any experience with mainframe data dictionaries know the power of such tools. They allow you to record facts about systems, both general and specific. They make it easier to track the impact of changes to specifications *before* you actually make the changes. And they make it simpler to automatically effect changes in many dictionary locations. For instance, you can change a data flow *label* and have that change automatically reflected in all data flow diagrams that include that data flow. If you change the size of a data element, the size for all data records that contain that element will be changed.

Before you learn the full power of EXCELERATOR's data dictionary (XLDICTIONARY), you need to learn some important terminology and technology. That is the purpose of this exercise. The term *XLDICTIONARY* is actually used to describe to components of EXCELERATOR, the dictionary itself, and the facility used to access and maintain the dictionary. This distinction is illustrated in Figure 7-1. In this lesson, you will learn how to use the facility; however, you must first understand the dictionary's composition and structure. Study the italicized terminology carefully. It will be used throughout Lessons 7, 8, and 9.

As Figure 7-1 illustrates, the data dictionary stores facts about your system project as you use all of the other facilities. To some extent, those facilities can also be used to create, maintain, and inspect the dictionary. However, certain types of maintenance can only be accomplished through the **XLDICTIONARY** facility, the subject of this lesson. You built a sizable dictionary in Part Two of this tutorial. Even the graphs are cataloged in the dictionary. Now, to make full use of the dictionary *facility*, you must study the dictionary's organization.

The dictionary stores occurrences of *entities* and *relationships*. The overall structure of the dictionary is illustrated in Figure 7-2. Rectangles represent *entity types*. Notice that there is an entity type for very type of graph, object, etc. in a system. EXCELERATOR provides each entity type with a standard set of *attributes* that can be recorded about any given occurrence of that entity type. For instance, almost all entity types have a free format attribute called **Description**. That attribute can be used to add a narrative description to any graph or object (Figure 7-3). In Figure 7-3, we have added a description attribute to an occurrence of the entity type, `Data Flow Diagram`. Figure 7-3 is an example of a Data Dictionary Description Screen. You saw several of these description screens in earlier lessons. Entity types have standard abbreviations (e.g. *DFD* for data flow diagram, *ERA* for entity-relationship diagram, *REC* for record, ELE for data element, and so forth).

Return to Figure 7-2. The named arrows depict *relationship types* that can exist between occurrences of entity types. Relationships are the most powerful concept in the dictionary. Every relationship type is named (e.g., *Contains*). In reality, the different relationship names are shared (e.g., you can see several instances where *contains* is a valid relationship type between different entity types). EXCELERATOR calls relationship types that have the same name *relationship classes*. The distinction between these important terms is summarized as follows:

o   a *relationship class* is a specific type of relationship (e.g., *explodes to, contains*).

o   a *relationship type* is a specific type of relationship that can exist between specific entities (e.g., process *explodes-to* data flow diagram, record *contains* data element).

123

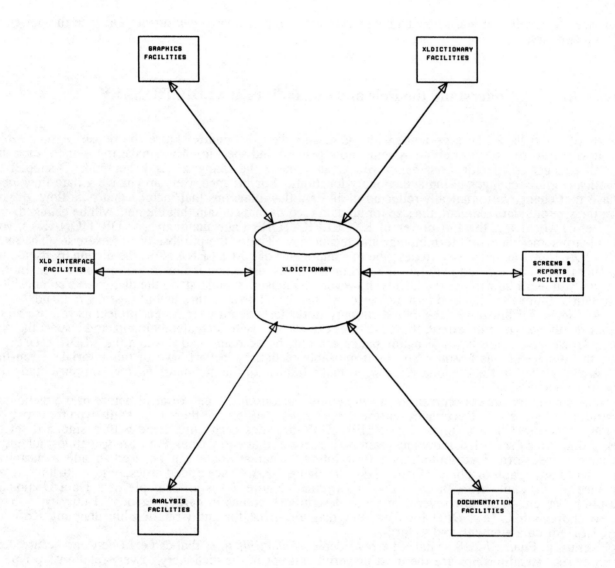

Figure 7-1

The Data Dictionary

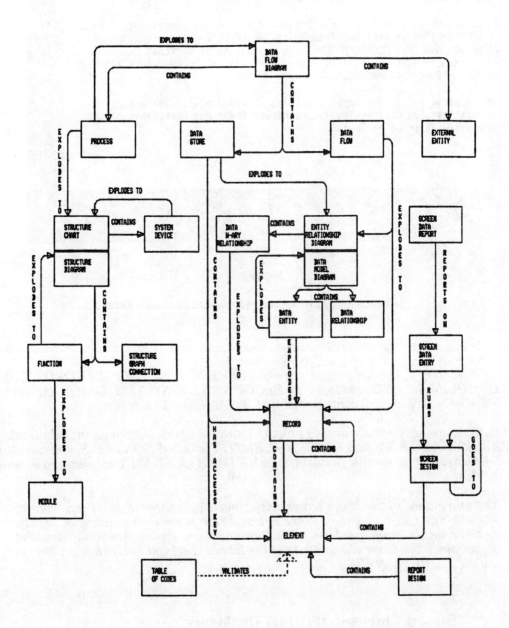

Figure 7-2

Data Dictionary Structure

125

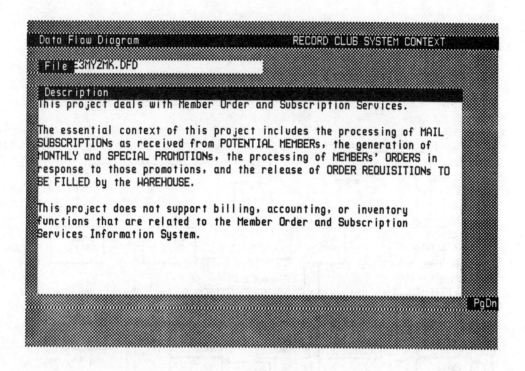

Figure 7-3

The Description Attribute of a Typical Data Dictionary Entity

o   a *relationship* is a specific occurrence of a relationship type (e.g., *RECORD CLUB CONTEXT DIAGRAM* (a DFD) *explodes to RECORD CLUB SYSTEM DIAGRAM* (another DFD), *MEMBER* (a RECord) *contains MEMBER NAME* (an ELEment).

Most relationships are bi-directional. Look at Figure 7-2 again. Although the relationship type between *DATA FLOW DIAGRAM* and *EXTERNAL ENTITY* is called *CONTAINS*, you can also use the dictionary to determine that an specific occurrence of EXTERNAL ENTITY is *contained in* which DATA FLOW DIAGRAMs.

> *The importance of the above terminology cannot be understated. If you learn it, you will have taken a big step toward unlocking the power of the dictionary facility. At this time, we recommend you reread this exercise to ensure that you understand the terminology. Once you understand the terminology, you ready to learn how to use the XLDICTIONARY facility.*

Exercise 7.2   Inspect Entries in the Data Dictionary

Now you are ready to learn how to use the **XLDICTIONARY** *facility*. This exercise will test your understanding of the terminology introduced in the last exercise. The easiest thing to learn how to do is to look up entries that you have already recorded in your dictionary (from Part Two of the tutorial). The first part of the exercise deals with looking up occurrences of an *entity type*. The second part deals with looking up occurrences of *relationship classes* and *types* (and why you want to do it).

Step 1.   Log on to EXCELERATOR and **Restore** your project directory. SELECT **XLDICTIONARY** from the Main Menu. This displays a menu of entity choices (Figure 7-4). Each choice provides a path to a submenu that includes one or more entity types as follows:

     o   **REC/ELE.** This option provides access to the entity types, Record and Element.

     o   **DATA.** This option provides access to the entity types, **Data Store, Data Entity, Data Flow, Table of Codes, Data Relationship,** and **Data N-Ary Relationship.** Some of these terms may be new to you - don't worry about them.

     o   **PROCESS.** This option provides access to the entity types **Process, Function, System Device, External Entity, Module, Presentation Graph Object, Structure Graph Connection,** and **Presentation Graph Connection.** Again, some of these terms may be new to you - don't worry about them.

     o   **GRAPHS.** This option provides access to the entity types **Data Flow Diagram, Structure Chart, Data Model Diagram, Entity-Relationship Diagram, Structure Diagram, Presentation Graph,** and **Document Graph.** Some of these graphs may be new to you - don't worry about them.

     o   **SCR/REPS.** This option provides access to the entity types **Report Design, Screen Design, Screen Data Entry,** and **Screen Data Reporting.** These options will only be familiar to you if you completed Lesson 6.

     o   **OTHER.** This option provides access to the entity types **Document Group, Document Fragment, Report, Entity List,** and **User,** most of which will be covered in Lessons 8 and 9.

Step 2.   Since all readers did Lesson 3, let's choose an option that includes specifications from that lesson. SELECT **DATA.** A submenu (Figure 7-5) appears. SELECT **Data Flow** as the entity type. Another submenu plus an Action Keypad appear (Figure 7-6). For the time being, ignore the second submenu (the one that begins with *1 DAF Contained-in DFD*).

Step 3.   You have several Action Keypad choices. First, SELECT **LIST.** The system prompts you for a **Name Range.** Let's just request a list of all the data flows currently in the dictionary. Press the Enter key.[1] You see a listing that looks something like Figure 7-7 (yours may vary depending on how many of the prior lessons you've completed and to what degree you completed them). To return to the Action Keypad, press the Esc key.

Step 4.   Next, SELECT **Inspect** from the Action Keypad. You are prompted for a name (actually an ID). Chances are, you won't remember all of your data flow IDs. Don't worry! Just press the Enter key. You will see a Selector List similar to those that you used in Part Two of the tutorial. Use the mouse to SELECT any name off the Selector List. This takes you to a Dictionary Description Screen similar to Figure 7-8.

This screen records up to four attributes: **Label** (which you have used), **Explodes to one of** (which you probably used in Part Two), **Duration Value** (not important at this time), **Duration Type** (again, not important at this time).

There are more attributes than the screen is currently displaying. Press any non-arrow key to display the next page of the data flow entry. This takes you to a screen similar to Figure 7-9. This screen has the additional attributes **Description** (a scroll region into which you can type any pertinent details) and several audit attributes that keep track of changes that have been made to this entity. (We'll learn more about these later.)

Press the PgUp key to return to Page 1 of the definition. Press any non-arrow key to return to the Action Keypad for **XLDICTIONARY.**

---

1   As a reminder, beginning with Version 1.7, the F3 function key can be frequently used in most cases where we suggest the use of the Enter key. However, to simplify the tutorial, we have adopted the convention of using the F3 function key only to save and exit data dictionary screens.

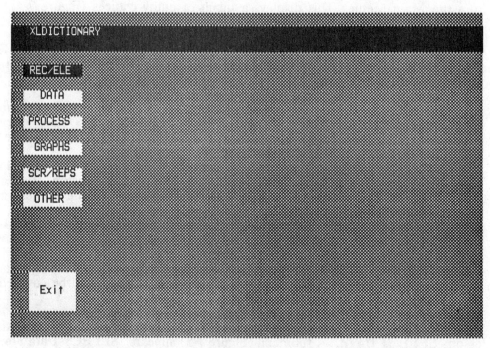

Figure 7-4

Classes of Dictionary Entities Accessible From the Dictionary

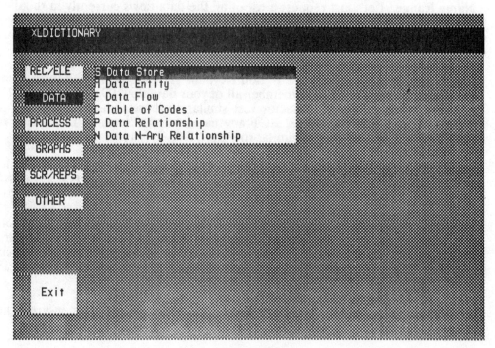

Figure 7-5

List of Entity Types for the Entity Class "DATA"

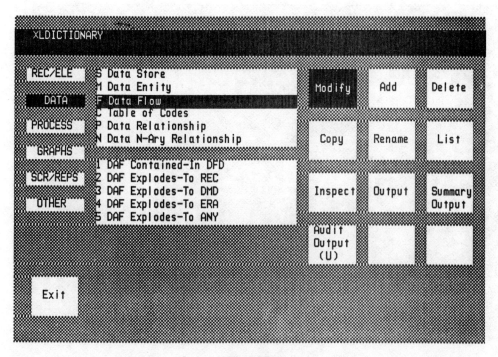

Figure 7-6

Action Keypad and Relationships Menu for "Data Flows"

```
XLDICTIONARY

REC/ELE      S Data Store
             M Data Entity              Modify    Add     Delete
  DATA       F Data Flow
             C Table of Codes
 PROCESS     P Data Relationship
             N Data N-Ary Relationship  Copy    Rename    List
  GRAPHS
             1 DAF Contained-In DFD
 SCR/REPS    2 DAF Explodes-To REC
             3 DAF Explodes-To DMD      Inspect Output   Summary
  OTHER      4 DAF Explodes-To ERA                       Output
             5 DAF Explodes-To ANY
                                        Audit
                                        Output
                                        (U)

  Exit
```

```
Data Flow Name                    Label
APPL: PROC APPLIC                 PROCESSED    APPLICATION
APPL: PROC APPLIC STATUS          PROCESSED    APPLICATION
APPL: PROSPECT MEM APPLIC         PROSPECTIVE  MEMBER       APPLICATION A
APPL: RENEW APPLIC                RENEWED      APPLICATION
BONUS: DEL BON REF                DELAY OF     BONUS FOR    REFERRAL
CAT: CAT ADV FLY PROMO            CATALOG AND  ADVERTISING  FLYERS TO BEI
INV: INV QUERY RESP               INVENTORY    CHECK ON-    LINE QUERY  R
LET: WEL/DENIAL LET               WELCOME OR   DENIAL       FORM LETTER
MEM: CUR MEM STATUS               CURRENT      MEMBER       STATUS
MEM: CURRENT MEM STATUS           CURRENT      MEMBER       STATUS
MEM: MEM DET                      MEMBER       DETAILS
MEM: MEM DET MUS PREF             MEMBER DE-   TAILS SORTEDBY MUSIC      P
MEM: MEM MUSIC PREF               MEMBER AND   MEMBER       MUSIC        P
MEM: NEW MEM DET                  NEW MEMBER   DETAILS
MEM: NOTIFI INELIBIBIL            NOTIFICATIONOF INELIGI-BILITY
MEM: POSS PAST DUE ACCT           POSSIBLE     PAST DUE     ACCOUNT
MEM: STAND TIME ACCT CLOSED       STANDING AT  TIME ACCOUNTWAS CLOSED
ORD: AUTO ORD FILL DATE PROMO     AUTOMATIC    ORDER FILL   DATE FOR     P
ORD: AUTO ORD FILLED              AUTOMATIC    ORDER TO BE FILLED
ORD: BACKORDER                    BACKORDER
ORD: BON COUP SPEC                BONUS        COUPONS      AND          S
Press any non-arrow key to see next screenful, [CANCEL] to return to menu
```

Figure 7-7

Selector List for Data Flows

129

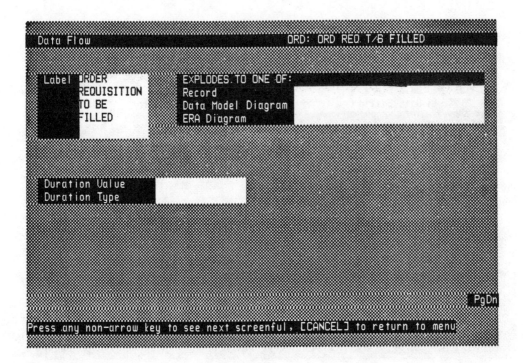

Figure 7-8

Data Flow Description Screen

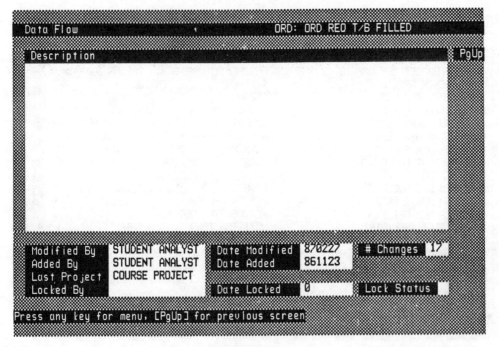

Figure 7-9

Second Page of Data Flow Description Screen

List and **Inspect** are available at any time if you need to check on any specific entity occurrence. Suppose, however, that you have a very large Selector List for the entity type that you wish to inspect. Selector lists can be several pages long for entity types such as data flow, data element, process, and record. When EXCELERATOR prompts you for a name, you can include wildcards in the response. Let's try it.

Step 5.   SELECT **List**. Type *MEM**. Press the Enter key. Depending on which lessons you've done, you should get a screen that looks something like Figure 7-10. Notice that only data flows that begin with *MEM* are listed. The IDs (names) are on the left. Your IDs will probably differ from ours. The labels are on the right. They should closely match your labels.

The wildcard asterisk must be the *last* character in the name string.

So now you know how to inspect entities. You probably expected that capability. The real inspection power lies in the ability to inspect relationships. First, you need to appreciate why you would want to inspect relationships. One of the great difficulties in maintaining system specifications is the problems caused when changes to one specification affect many other specifications. Relationships allow you to preview the possible impact of such changes.

For example, suppose that you decide to change the size of a data element. Think about it! It could be a dramatic change. The size of records that contain that element will change. Reports and screen designs containing that element will have to be altered. Validation rules may have to be changed. In other words, that one little change could cause many other little changes, each of which could cause major problems if the change and impact are not reflected throughout the specifications.

Using **XLDICTIONARY**, you can look up relationships before recording changes (changes are covered in the next exercise). Here's how. Since we don't know which lessons you covered in Part Two, we must offer two options for teaching you this technique. The options are noted by the letters A (use this one if you completed Lesson 5, data modeling) and B (use this one if you only completed Lesson 4, data flow modeling) after the step numbers. Option A is preferred.

Option A

Step 5A.   SELECT **REC/ELE** from the left-hand menu. SELECT **Element** when the submenu appears. Once again, both another submenu and the Action Keypad appear. The second submenu describes the relationships tracked by EXCELERATOR for the entity type **Element**. There are four relationships:

o   **ELE Contained-In REC.** Indicates that a data element may be contained in record layouts.
o   **ELE Contained-In SCD.** Indicates that a data element may be contained in screen designs.
o   **ELE Contained-In RED.** Indicates that a data element may be contained in report designs.
o   **ELE Access-Key-Of DAS.** Indicates that a data element may be the key of data stores.

The abbreviations are described more thoroughly in InTech's Reference Manual.

Let's SELECT **ELE Contained-In REC**. The Action Keypad changes (Figure 7-11) to reflect those actions appropriate to the selections you have made.

Step 6A.   SELECT **List**. The system prompts you for a name range. You can use wildcards here; however, you should just press the Enter key this time. A report similar to Figure 7-12 should appear. It shows precisely which elements are contained in which records. A change to any element would affect all of those records. Press any non-arrow key to exit the report.

You should use the above capability to track the possible effect of changes before you make any changes. You can then inspect any of the affected entities prior to making the changes. You can now proceed to Exercise 7.3 (unless you want to try Option B for additional practice).

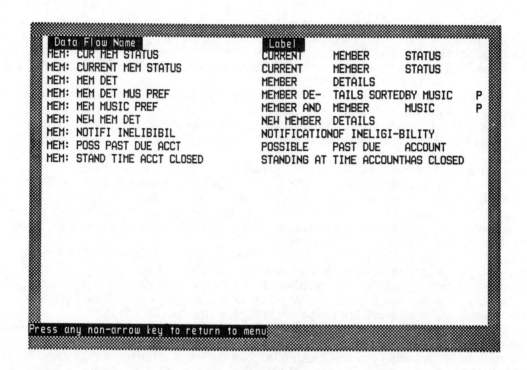

```
 Data Flow Name                         Label
MEM: CUR MEM STATUS            CURRENT     MEMBER      STATUS
MEM: CURRENT MEM STATUS        CURRENT     MEMBER      STATUS
MEM: MEM DET                   MEMBER      DETAILS
MEM: MEM DET MUS PREF          MEMBER DE-  TAILS SORTEDBY MUSIC    P
MEM: MEM MUSIC PREF            MEMBER AND  MEMBER      MUSIC       P
MEM: NEW MEM DET               NEW MEMBER  DETAILS
MEM: NOTIFI INELIBIBIL         NOTIFICATIONOF INELIGI-BILITY
MEM: POSS PAST DUE ACCT        POSSIBLE    PAST DUE    ACCOUNT
MEM: STAND TIME ACCT CLOSED    STANDING AT TIME ACCOUNTWAS CLOSED




Press any non-arrow key to return to menu
```

Figure 7-10

Use of Wildcard Names

```
 XLDICTIONARY


 REC/ELE    R Record
            E Element

   DATA
            1 ELE Contained-In REC
 PROCESS    2 ELE Contained-In SCD
            3 ELE Contained-In RED                      List
  GRAPHS    4 ELE Access-Key-Of DAS

 SCR/REPS
                                                      Summary
  OTHER                                               Output

                                        Missing
                                        Entities
                                        Output

 Exit
```

Figure 7-11

Typical Action Keypad for Examining Relationships Between Entity Types

132

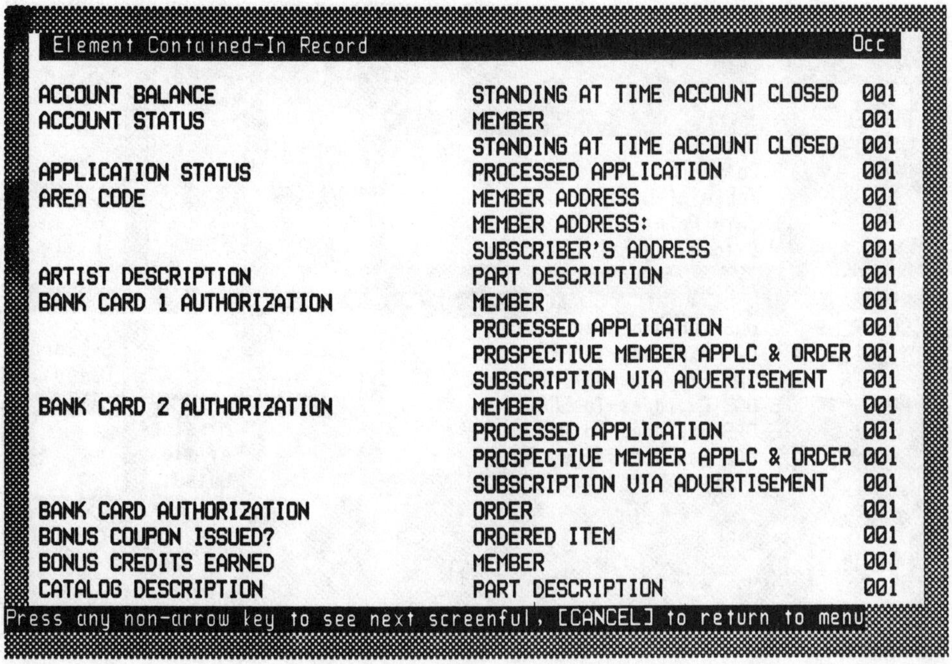

```
┌──────────────────────────────────────────────────────────────────────┐
│ Element Contained-In Record                                       Occ  │
├──────────────────────────────────────────────────────────────────────┤
│ ACCOUNT BALANCE          STANDING AT TIME ACCOUNT CLOSED          001  │
│ ACCOUNT STATUS           MEMBER                                   001  │
│                          STANDING AT TIME ACCOUNT CLOSED          001  │
│ APPLICATION STATUS       PROCESSED APPLICATION                    001  │
│ AREA CODE                MEMBER ADDRESS                           001  │
│                          MEMBER ADDRESS:                          001  │
│                          SUBSCRIBER'S ADDRESS                     001  │
│ ARTIST DESCRIPTION       PART DESCRIPTION                         001  │
│ BANK CARD 1 AUTHORIZATION MEMBER                                  001  │
│                          PROCESSED APPLICATION                    001  │
│                          PROSPECTIVE MEMBER APPLC & ORDER         001  │
│                          SUBSCRIPTION VIA ADVERTISEMENT           001  │
│ BANK CARD 2 AUTHORIZATION MEMBER                                  001  │
│                          PROCESSED APPLICATION                    001  │
│                          PROSPECTIVE MEMBER APPLC & ORDER         001  │
│                          SUBSCRIPTION VIA ADVERTISEMENT           001  │
│ BANK CARD AUTHORIZATION  ORDER                                    001  │
│ BONUS COUPON ISSUED?     ORDERED ITEM                            001  │
│ BONUS CREDITS EARNED     MEMBER                                   001  │
│ CATALOG DESCRIPTION      PART DESCRIPTION                         001  │
├──────────────────────────────────────────────────────────────────────┤
│ Press any non-arrow key to see next screenful. [CANCEL] to return to menu │
└──────────────────────────────────────────────────────────────────────┘
```

Figure 7-12

List Report for a Relationship Class

Option B

Step 5B. SELECT **DATA** from the left-hand menu. SELECT **Data Store** when the submenu appears. Once again, both another submenu and the Action Keypad appear. The second submenu describes the relationships tracked by EXCELERATOR for the entity type **Data Store**. They are described as follows:

o **DAS Contained-In DFD.** Indicates that a data store may be contained in data flow diagrams.

o **DAS Explodes-To REC.** Indicates that a data store may be exploded into record layouts.

o **DAS Explodes-To DMD.** Indicates that a data store may be exploded into Bachman data model diagrams.

o **DAS Explodes-To ERA.** Indicates that a data store may be exploded to an entity relationship data model diagram.

o **DAS Explodes-To ANY.** Indicates that data stores may be exploded to any combination of records, data model diagrams, and entity-relationship diagrams. This is useful for tracking all of the explosion paths with a single relationship class.

o **DAS Has-Access-Key ELE.** Indicates that data stores have elements as access keys.

The abbreviations are described more thoroughly in InTech's Reference Manual.

Let's SELECT **DAS Contained-In DFD.** The Action Keypad changes (Figure 7-13) to reflect those actions appropriate to the selections you have made.

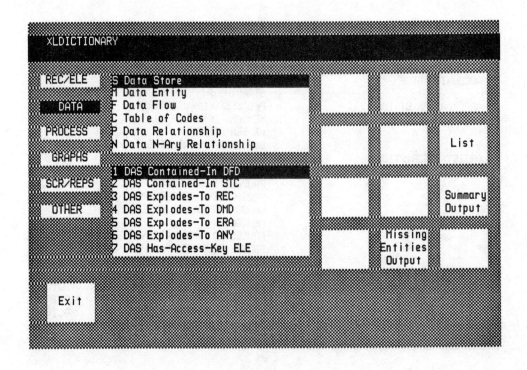

Figure 7-13

Typical Action Keypad for Examining Relationships Between Entity Types

Step 6B.   SELECT **List**. The system prompts you for a name range. You can use wildcards here; however, you should just press the Enter key this time. A report similar to Figure 7-14 should appear. It shows precisely which elements are contained in which records. A change to any element would affect all of those records. Press any non-arrow key to exit the report.

Individual names or wildcard lists could have been entered at the name range prompt.

You should use the above capability to track the possible effect of changes before you make any changes. You can then inspect any of the affected entities prior to making the changes. To conclude this exercise, SELECT **Exit**. This returns you to the Main Menu.

Exercise 7.3   Maintain Entries in the Data Dictionary

The **XLDICTIONARY** provides the capability of maintaining specifications that were created when using the **GRAPHICS** facility (Lessons 3 through 5). Why do you need another facility? Why not just use the **GRAPHICS** facility? There are two reasons. First, **XLDICTIONARY** lets you pre-determine the impact of your changes throughout your entire system specification. Second, there are certain types of descriptions and attributes that can be maintained only via the **XLDICTIONARY** facility. Let's add and change some of the descriptions that are already in your dictionary. We'll also explain how to perform deletions.

Step 1.   SELECT **XLDICTIONARY**. Then SELECT **GRAPHS**. Next, SELECT **Data Flow Diagram**.

Step 2.   Notice that the Action Keypad (Figure 7-15) contains no *Add* option. Graph descriptions are automatically added when you draw the actual graphs (via the **GRAPHICS** facility).   Now we

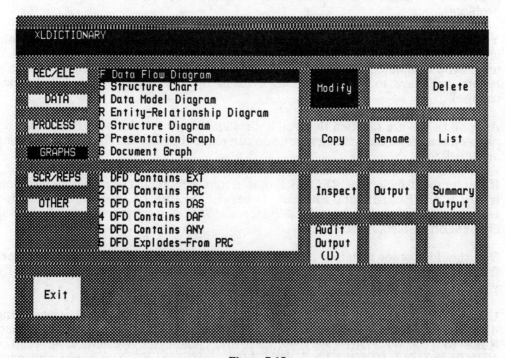

```
Data Store Contained-In Data Flow Diagram
APPL                              SUBSCRIPTION PROCESSING
                                  SUBSCRIPTION PROCESSING COPY
MEM                               AUTOMATIC ORDER PROCESSING
                                  BONUS/SPECIAL ORDER PROCESSING
                                  MAILED IN ORDER PROCESSING
                                  ORDER PROCESSING
                                  PROMOTION PROCESSING
                                  RECORD CLUB SYSTEM DIAGRAM
                                  SUBSCRIPTION PROCESSING
                                  SUBSCRIPTION PROCESSING 1ST DRFT
                                  SUBSCRIPTION PROCESSING COPY
MEMCR                             AUTOMATIC ORDER PROCESSING
                                  BONUS/SPECIAL ORDER PROCESSING
                                  MAILED IN ORDER PROCESSING
MEMP                              SUBSCRIPTION PROCESSING
                                  SUBSCRIPTION PROCESSING 1ST DRFT
                                  SUBSCRIPTION PROCESSING COPY
MEMSD                             PROMOTION PROCESSING
ORD                               AUTOMATIC ORDER PROCESSING
                                  BONUS/SPECIAL ORDER PROCESSING
Press any non-arrow key to see next screenful, [CANCEL] to return to menu
```

Figure 7-14

List Report for a Relationship Class

```
XLDICTIONARY

REC/ELE    F Data Flow Diagram
           S Structure Chart           Modify              Delete
DATA       M Data Model Diagram
           R Entity-Relationship Diagram
PROCESS    D Structure Diagram
           P Presentation Graph        Copy    Rename      List
GRAPHS     G Document Graph

SCR/REPS   1 DFD Contains EXT
           2 DFD Contains PRC          Inspect Output      Summary
OTHER      3 DFD Contains DAS                              Output
           4 DFD Contains DAF
           5 DFD Contains ANY          Audit
           6 DFD Explodes-From PRC     Output
                                       (U)

Exit
```

Figure 7-15

Action Keypad for Data Flow Diagrams

want to add details to those blank graph descriptions. Consequently, SELECT Modify. The system prompts you for a name. Press the Enter key to see a Selector List of names. Choose any name from your Selector List. A description screen similar to Figure 7-3 appears (although it is probably blank).

Step 3.   In Figure 7-16, we provide printouts of descriptions from all graphs drawn in Lessons 4. Type these descriptions in the **Description** block attribute of the screen. The **Description** block works like a word processor - that is, unlike **Labels**, **Description** blocks support word wrapping. If the word you are typing won't fit on the line, it will automatically wrap around to the next line. The only time you need to press the Enter Key is when you want to force the cursor to a new paragraph or line.[2]

Press the F3 function key to save your work and return to the Dictionary Menu.

So much for the simple change to an entity. What about something more substantial. Consider, for example, changing the size of a data element. At first glance, this may seem like a simple modification. But consider the ramifications! For instance, an increased element size changes the size of records, corresponding COBOL Data Divisions, report layouts, screen layouts, validation rules, etc. EXCELERATOR provides a means of checking the impact before recording the changes. Also, EXCELERATOR makes many changes throughout the specifications (note: data element changes do not affect corresponding fields in screen and report designs). Let's change *MEMBER NUMBER* in the record club case from a five digit number to a seven digit number.

Step 4.   First, let's determine the impact of the proposed change. SELECT **REC/ELE.** SELECT **Element.** SELECT **ELE Contained-In REC.** SELECT **List** from the Action Keypad. Type the name *MEMBER NUMBER* or SELECT that element from a Selector List. You'll see a report (similar to Figure 7-17). Jot down one or more names for later reference (we intend to demonstrate that the changes do ripple to these records). If you were to see potential problems, you could double check those records by **Inspect**ing them. Let's say that we see no problems.

You could (and normally would) repeat this step for the relationships **ELE Contained-In SCD** and **ELE Contained-In RED.** These relationships won't reveal any additional implications unless you completed Lesson 6 (on prototyping). This listing will tell you which screens and reports will have to be modified as a result of your change (they will not automatically change).

Step 5.   As soon as you are satisfied that you understand which entities will change as a result of your planned change to the data element, you can make the planned change. Let's assume that you are satisfied.

If you temporarily left the **REC/ELE** option to inspect report or screen descriptions, you must re-SELECT **REC/ELE** and **Element.** Otherwise, SELECT **Element** and then **Modify.** Either type the name of the element, *MEMBER NUMBER*, and press the Enter key, or SELECT the element off a Selector List.

Step 6.   You should now see a Data Element Description Screen (Figure 7-18). This screen should be familiar to you from Lessons 4 through 6. The attributes should be at least partially filled in (if you described this element during Lessons 4-6).

Change the **Characters left of decimal** to *7*. The new size also necessitates changes to other data element attributes. Change the **Input and Output Pictures** to *9999999*. Change the **Edit Rules** to *1000000 THRU 9999999*. Finally, change the Definition to read *A 7 DIGIT NUMBER THAT UNIQUELY IDENTIFIES A MEMBER*. You could also page down and type a detailed Description. To save the changes, press the F3 function key.

---

2   We remind you that you can also use the Tab key in most places where we suggest the Enter key. For versions earlier than 1.7, you must use the Tab key.

```
DATE: 27-MAR-87       DATA FLOW DIAGRAM - OUTPUT                          PAGE    6
TIME: 15:08           NAME: *                                      EXCELERATOR  1.7

TYPE Data Flow Diagram                    NAME RECORD CLUB SYSTEM CONTEXT

   File E3MYZMK.DFD

   Description
  This project deals with Member Order and Subscription Services.

  The essential context of this project includes the processing of MAIL
  SUBSCRIPTIONs as received from POTENTIAL MEMBERs, the generation of
  MONTHLY and SPECIAL PROMOTIONs, the processing of MEMBERs' ORDERS in
  response to those promotions, and the release of ORDER REQUISITIONs TO
  BE FILLED by the WAREHOUSE.

  This project does not support billing, accounting, or inventory
  functions that are related to the Member Order and Subscription
  Services Information System.


DATE: 27-MAR-87       DATA FLOW DIAGRAM - OUTPUT                          PAGE    7
TIME: 15:08           NAME: *                                      EXCELERATOR  1.7

TYPE Data Flow Diagram                    NAME RECORD CLUB SYSTEM DIAGRAM

   File E3N2ZY8.DFD

   Description
  The RECORD CLUB SYSTEM is composed of three subsystems:  SUBSCRIPTION
  SUBSYSTEM, PROMOTION SUBSYSTEM, and ORDER PROCESSING SUBSYSTEM.  This
  data flow diagram depicts these three subsystems and their interfaces.

  Essentially, the SUBSCRIPTION SUBSYSTEM is responsible for processing
  all MAIL SUBSCRIPTIONS received from POTENTIAL MEMBERs.  The PROMOTION
  SUBSYSTEM, in response to RECORD PROMOTIONS received from the MARKETING
  DEPARTMENT, generates MONTHLY OR SPECIAL PROMOTIONS for notifying
  current club MEMBERS.  Finally, the ORDER PROCESSING SUBSYSTEM
  processes MEMBER's ORDERS (resulting from promotions) and releases
  ORDER REQUISITIONS TO BE FILLED by the WAREHOUSE.


DATE: 27-MAR-87       DATA FLOW DIAGRAM - OUTPUT                          PAGE    8
TIME: 15:08           NAME: *                                      EXCELERATOR  1.7

TYPE Data Flow Diagram                    NAME RECORD CLUB SYSTEM OVERVIEW

   File E3PIYDI.DFD

   Description
  This project deals with Member Order and Subscription Services.

  This graph depicts the structure of the system.  The RECORD CLUB SYSTEM
  can conveniently be broken down into three subsystems including:

        1.   SUBSCRIPTION SUBSYSTEM - This subsystem involves the
             processing of club membership MAIL SUBSCRIPTIONS

        2.   PROMOTION SUBSYSTEM - This subsystem involves the generation
             of MONTHLY OR SPECIAL PROMOTIONS.

        3.   ORDER PROCESSING SUBSYSTEM - This subsystem involves
             processing orders and releasing ORDER REQUISITIONS TO BE FILLED
             by the WAREHOUSE.  It can further be broken down into three
             functional areas dedicated to processing specific order types.


DATE: 27-MAR-87       DATA FLOW DIAGRAM - OUTPUT                          PAGE    9
TIME: 15:08           NAME: *                                      EXCELERATOR  1.7

TYPE Data Flow Diagram                    NAME SUBSCRIPTION PROCESSING

   File E3M21XB.DFD

   Description
  This data flow diagram details the SUBSCRIPTION SUBSYSTEM of the RECORD
  CLUB SYSTEM.

  The SUBSCRIPTION SUBSYSTEM involves the verification of SUBSCRIPTIONs
  VIA REFERRAL, the approval and notification of the applicant (POTENTIAL
  MEMBER), and ORDER transcription.
```

Figure 7-16(a)

Data Flow Diagram Descriptions (Part 1)

DATE: 27-MAR-87 DATA FLOW DIAGRAM - OUTPUT PAGE 1
TIME: 15:08 NAME: * EXCELERATOR 1.7

TYPE Data Flow Diagram NAME AUTOMATIC ORDER PROCESSING

File E3OU2JP.DFD

Description
This data flow diagram depicts the data flow and processing of automatic orders.

AUTOMATIC ORDER PROCESSING is triggered by the PROMOTION SUBSYSTEM when notice of the AUTOMATIC ORDER FILL DATE FOR PROMOTION is received. This date is used to identify orders that should be generated because club members elected not to respond to a MONTHLY OR SPECIAL PROMOTION offering. Those orders are identified and ORDER REQUISITIONS TO BE FILLED are generated and sent to the WAREHOUSE.

DATE: 27-MAR-87 DATA FLOW DIAGRAM - OUTPUT PAGE 2
TIME: 15:08 NAME: * EXCELERATOR 1.7

TYPE Data Flow Diagram NAME BONUS/SPECIAL ORDER PROCESSING

File E3PM20T.DFD

Description
This data flow diagram represents the processing of SPECIAL ORDER AND/OR BONUS COUPONS received from MEMBERs.

Essentially, SPECIAL ORDER AND/OR BONUS COUPONS are screened to make sure that a MEMBER is eligible for filling such an order. The MEMBER's credit standing and inventory levels are also checked prior to releasing an ORDER REQUISITION TO BE FILLED by the WAREHOUSE.

DATE: 27-MAR-87 DATA FLOW DIAGRAM - OUTPUT PAGE 4
TIME: 15:08 NAME: * EXCELERATOR 1.7

TYPE Data Flow Diagram NAME ORDER PROCESSING

File E3OY8YJ.DFD

Description
This data flow diagram depicts the ORDER PROCESSING SUBSYSTEM. This subsystem involves the processing of automatic orders, mail orders, and bonus or special orders. The ORDER PROCESSING SUBSYSTEM is conveniently broken down into functional three functional areas responsible for processing a given order type. These three areas and there interfaces are depicted.

DATE: 27-MAR-87 DATA FLOW DIAGRAM - OUTPUT PAGE 5
TIME: 15:08 NAME: * EXCELERATOR 1.7

TYPE Data Flow Diagram NAME PROMOTION PROCESSING

File E3MQ0M0.DFD

Description
This data flow diagram depicts the PROMOTION SUBSYSTEM of the RECORD CLUB SYSTEM. This subsystem is responsible for the generation of MONTHLY OR SPECIAL PROMOTIONS.

Figure 7-16(b)

Data Flow Diagram Descriptions (Part 2)

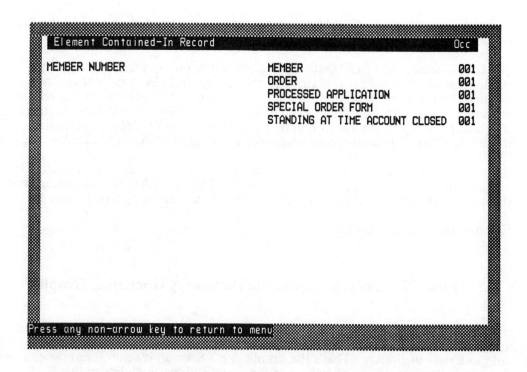

```
Element Contained-In Record                              Occ

MEMBER NUMBER                    MEMBER                     001
                                ORDER                      001
                                PROCESSED APPLICATION      001
                                SPECIAL ORDER FORM         001
                                STANDING AT TIME ACCOUNT CLOSED  001











Press any non-arrow key to return to menu
```

Figure 7-17

Records Containing the Data Element MEMBER NUMBER

```
Element                              MEMBER NUMBER

  Alternate Names  MEMBERSHIP NUMBER



  Definition       A 5 DIGIT NUMBER THAT UNIQUELY IDENTIFIES A MEMBER

  Input Picture    99999
  Output Pic       99999
  Edit Rules       10000 THRU 99999

  Storage Type     I
  Characters left of decimal          Characters right of decimal  0
  Prompt           MEMBERSHIP NUMBER?
  Column Header    MEMBER NUMBER
  Short Header     MEMBER#
  Base or Derived  D
  Data Class
  Source           SUBSCRIPTION MANAGER
  Default                                                       PgDn
```

Figure 7-18

Data Element Description Screen

139

Step 7.   Just to be sure that the effect of our change is now reflected elsewhere in the dictionary, let's check one of the records that contain the element.

SELECT **Record**. SELECT **Output**. We chose the output function because it displays both the lengths of all elements contained in the record and also the total record length. When prompted for a name, type (or SELECT from a Selector List) one of the record names you just jotted down. Press the Enter key. SELECT either **Screen** or **Printer** as the output device. You will get a report similar to Figure 7-19. Notice that *MEMBER NUMBER*'s LENgth has indeed changed. The total record length has already been adjusted. Slick!

For your information, EXCELERATOR does *not* automatically change fields on the prototypes. However, you can return to the **SCREENS & REPORTS** functions and easily use the new *MEMBER NUMBER* element to adjust associated fields (see Lesson 6).

Step 8.   Exit to return to the Main Menu.

Exercise 7.4   Track Effects of Changes to the Dictionary (Including Graphs)

You've already seen that changes made to one entity are automatically recorded in other entities. This exercise serves two additional purposes. First, it demonstrates how EXCELERATOR audits and reports the changes made to a data dictionary. That's the simple part. Second, it reinforces how changes made in the dictionary are automatically updated elsewhere in the specifications, including graphs!

One of the luxuries afforded by **XLDICTIONARY** is that it automatically keeps track of who makes changes to the dictionary. None of these audit attributes are directly modifiable; however, you should be aware of their existence and how to read and use them.

Step 1.   From the Main Menu, SELECT **XLDICTIONARY**. SELECT **DATA**. SELECT **Data Flow**. SELECT **Modify** from the Action Keypad. Press the Enter key to see a Selector List of data flow entities. Jot down the ID (short name on the left-hand side of the screen) of the data flow labeled *MAIL SUBSCRIPTION*. SELECT that data flow from the list. (Note: the data flow originally appeared on Figure 3-6.)

Step 2.   You should now be at a dictionary description screen. Press the PgDn key go see the second page of the entry (Figure 7-20). The audit attributes are at the bottom of the screen:

o   **Modified By.** Indicates the user account name that last modified this entity.
o   **Added By.** Indicates the user account name that first created the entity.
o   **Last Project.** Indicates the project name that last modified the entity.
o   **Locked By.**   Locking prevents unauthorized or accidental modifications or deletions. If locked, this attribute indicates which user account name did the locking (especially important if dictionaries are to be shared between users or projects). Locking is covered later in the lesson.
o   **Date Modified.** Indicates the date of the last modification to the entity.
o   **Date Added.** Indicates the date on which the entity was first created in the dictionary.
o   **Date Locked.** Indicates the date on which the entity was locked.
o   **# Changes.** Indicates how many changes have been made to the description screens.
o   **Date Locked.** Indicates the date on which the entity was locked.
o   **Lock Status.**   Indicates whether the entity is currently locked to prevent unwanted changes.

These attributes can help you track changes, even when dictionaries are shared by multiple users and projects.

The next part of this lesson reinforces the completeness relative to the impact of a change in to any entity recorded in the dictionary. We want you to change a graphics data flow Label to demonstrate that the change will also be reflected in all graphs that contain the data flow.

```
DATE: 27-MAR-87                          RECORD - EXPLOSION
TIME: 15:16                              NAME: MEMBER

NAME:                          MEMBER                          DEFINITION:
ALIAS:

ELEMENT/RECORD                                OFF  OCC  TYPE  LEN  DEFINITION
----------------------------------------      ---  ---  ----  ---  ----------

MEMBER NUMBER                                 000  001   1    007  A 7 DIGIT NUMBER THAT U
MEMBER NAME                                   007  001   2    030  LAST NAME, FIRST NAME
DATE ENROLLED                                 037  001   E    002  DATE THAT MEMBER FIRST
BANK CARD 1 AUTHORIZATION                     039  001   E    022  BANK CARD NUMBER TO BE
BANK CARD 2 AUTHORIZATION                     061  001   E    022  BANK CARD NUMBER TO BE
MEMBER ADDRESS:                               083  001   R
   STREET ADDRESS                             083  001   9
   POST OFFICE BOX NUMBER                     083  001   9
   CITY                                       083  001   E
   STATE                                      098  001   E    015  CITY OF MEMBER'S RESIDE
   ZIP CODE                                   098  001   E    002  A TWO CHARACTER CODE OF
   AREA CODE                                  100  001   E
   PHONE NUMBER                               100  001   E
                                              100  001   E
MUSICAL PREFERENCE                            100  001   8    002  CATEGORY OF MUSIC PREFE
MEDIUM PREFERENCE                             102  001   8    002  DEFAULT RECORDING MEDIU
MEMBER BALANCE DUE                            104  001   E    003  UNPAID BALANCE DUE ON M
ACCOUNT STATUS                                107  001   8    001  MEMBER'S ACCOUNT STATUS
MEMBERSHIP PURCHASE REQUIREMENT               108  001   E    002  THE NUMBER OF PURCHASES
MEMBERSHIP EXPIRATION DATE                    110  001   8    002  DATE BY WHICH MINIMUM R
MEMBER PURCHASE CREDITS EARNED                112  001   E    002  NUMBER OF PURCHASES THA
MEMBER PURCHASES BEYOND REQT                  114  001   E    002  PURCHASES BEYOND MINIMU
BONUS CREDITS EARNED                          116  001   E    002  BONUS COUPONS THAT HAVE
Record length is 118.
```

Figure 7-19

Printout of a Record Description

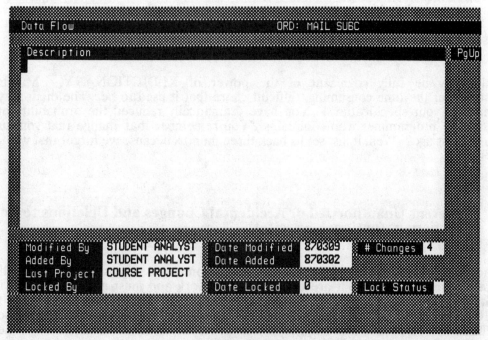

Figure 7-20

Audit Attributes of a Data Flow Description Screen

141

Step 3.   Press the PgUp key to return to the first description screen. We have decided to more accurately label the data flow to reflect that the data flow includes a first order. The cursor should already be at the **Label** attribute; therefore, modify that attribute as shown in Figure 7-21 (simply type over the original label). Press the F3 function key to save your work and return to the Action Keypad.

Step 4.   We want to check the DFDs to convince ourselves that the dictionary is using known relationships to propagate changes to the actual graphs. First, we need to know what graphs were changed. That part is easy. SELECT **DAF Contained-In DFD**. SELECT List. Type the name and press the Enter key. (Important: Selector Lists are not available from the **List**, **Summary Output**, or **Audit Attribute** commands.) You should now see a list of data flow diagrams that contain the data flow; the list size will depend on how many lessons you fully completed in Part Two. Jot down the names of at least one of those data flow diagrams. Press any non-arrow key to return to the Action Keypad for **XLDICTIONARY**.

Step 5.   To see the impact of our change, we must get into the **GRAPHICS** facility. SELECT Exit. This takes you back to the Main Menu. SELECT **GRAPHICS**. SELECT **Data Flow Diagram**. SELECT **Modify**. Press the Enter key in response to the name prompt. SELECT one of the names that you jotted down in Step 4.

Step 6.   If necessary, **ZOOM** to **CLOSE UP**. Pan to the data flow we changed. Notice that the label has already been changed to reflect the new name. You don't have to change each graph, or even remember to change them. If you are still not convinced, EXIT the graph and check another DFD that contains the changed data flow. We guarantee that all data flows have been changed!

We would be remiss if we didn't point out that this change worked throughout the DFDs because we changed the *label*, not the ID (Name). If you changed the name of a data flow, you *must* correct each individual occurrence of the data flow (DESCRIBE each with the new ID). Consequently, change data flow IDs only if necessary. If the user objects to the label, change only the label.

Step 7.   As soon as you are convinced, **EXIT** to the Main Menu.

You should now be fully cognizant of the power of XLDICTIONARY. Maintenance of specifications is no longer the time consuming, difficult chore that it used to be. The dictionary increases your own confidence in your specifications! You have dramatically reduced the probability of the once inevitable visit from the programmer who complains, "You remember that change that you made to the specifications two months ago? Well, it just set us back three months because we forgot that we have to also change ..."

Exercise 7.5   **Prevent Unauthorized or Accidental Changes and Deletions to Your Dictionary (optional exercise)**

This exercise is not appropriate to all EXCELERATOR environments. The situation is as follows. You have one or more entities or graphs that are critical to your work and must not be altered without very careful consideration or approval. Or perhaps you are lending all or part of your dictionary to another EXCELERATOR user account or project (possibly a subproject of your own). How do you prevent accidental or unauthorized changes to an entity(ies)? The procedure is *not* handled via the dictionary. It is called *locking* and *unlocking* an entity(ies).

Step 1.   From the Main Menu, SELECT **XLD INTERFACE**. This takes you to a new menu (Figure 7-22). SELECT **Lock**. SELECT **Via SELECTION**. You can select entities by **Type** (e.g., ELE, DFD, or REC). You can select entities by name, including wildcards. And you can select entities by who last modified them.

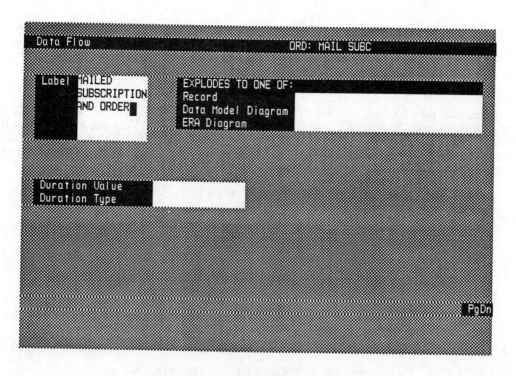

Figure 7-21

Modified Data Flow Description Screen

DICTIONARY INTERFACE

1 Export

2 Import

3 Lock

4 Unlock

5 Export & Lock

6 Import & Unlock

Exit

Figure 7-22

XLD INTERFACE Submenu Screen

SELECT **Type.** Type *DFD*. This will lock entities of the type *data flow diagram*. You can specify up to five selection criteria; however, we want to lock all DFDs. Therefore, SELECT **Execute** to complete the task.

The selected entities are now locked. Other accounts can inspect their contents but not change them. Entities must be unlocked from another ID with the same access privilege. You can easily check the results of your last action. Simply enter the **XLDICTIONARY** facility, SELECT **DATA**, SELECT **Data Flow Diagram**, SELECT any DFD by name, page down (PgDn key), and study the audit attributes for that entity. They will show that the entity is indeed locked, that you did it, and that you did it moments ago.

To unlock the entities follow this procedure.

Step 2.   SELECT **Unlock.** SELECT **via SELECTION.** Once again, the list of selection fields will appear. SELECT **Execute.**

Locking and unlocking are especially useful for controlling change in multiple-analyst projects. It can also be used to share templates. Templates are skeleton graphs or dictionary standards that other analysts can use as models or starting points in other projects.

Step 3.   SELECT **Exit.** This returns you to EXCELERATOR's Main Menu.

Exercise 7.6   Output the Data Dictionary

This is a simple exercise. The purpose is to teach you how to generate a wide variety of outputs from the **XLDICTIONARY** facility. Outputs differ in level of content and number of entities and relationships reported. You can complete this exercise whether or not you have a printer. Outputs can be directed to printers, screens, or files (which can then be printed on other microcomputers). 132 column reports will wrap on your 80 column display monitor.

Step 1.   From the Main Menu, SELECT **XLDICTIONARY.** SELECT **DATA.** SELECT **Data Flow.** There are several output options on the Action Keypad. They are described as follows:

o   **List.** You've already used this option. List displays only IDs and labels.

o   **Inspect.** You've also used this option. Inspect retrieves specific entity occurrences for display. You cannot modify any attributes from the Inspect option. You also cannot get a hardcopy printout from the option.

o   **Output.** This is the main output option. It prints the entire dictionary contents for selected entity names or name ranges. All attributes are output, including the audit attributes. Output can be directed to the a file, the screen, or a printer.

o   **Summary Output.** This output prints the name (ID) and two other attributes for the selected entity names or name range. The contents vary depending on the entity type selected. Consequently, it is best to initially route the output to the screen before routing it to a printer or a file. This way, you can ensure that the attributes displayed are the ones you want.

o   **Audit Output (U).** This output prints only the audit attributes for the selected entity name or name range.

A few explanations are in order. All output options prompt you for a name or name range. You can only call up Selector Lists for the Inspect and Output options. For the other options, you must give a name or name range. Wildcards, as usual, are permitted.

For the **Output** option only, you are asked if you want output fields underlined. By this, EXCELERATOR means that it will not underline attribute headings but will underline the values that you entered for those attribute headings. We recommend that you respond with Y for "yes." Otherwise, you may confuse attribute headings and values.

Finally, EXCELERATOR gives you three options for output device. If you SELECT **Printer**, you will get a hardcopy printout (assuming you have a printer). If you SELECT **Screen**, the output will be displayed on the screen. You can pause the scrolling by simultaneously pressing the Ctrl and S keys. Finally, you can SELECT the left most option, a **.PRN** print file name. EXCELERATOR prompts you to change the file name. If you press the Enter key, the name displayed will be used. The file can eventually be copied to diskette, loaded on another machine, and printed via DOS's standard *PRINT* command.

Step 2.   When prompted for a name or name range, you can type a name (including wildcards). For this exercise, press the Enter key to see a Selector List. Notice the first entry reads *All entities on Selector List*. This option, if selected, would print all data flow entries. for all data flows. Let's not select that option. Instead, SELECT *SUBSCRIPTION VIA ADVERTISEMENT*. Press Y to indicate that you want output fields underlined. Next, SELECT either **Screen** or **Printer**. Your output should be similar to Figure 7-23.

Step 3.   That completes the exercise and lesson. SELECT **Exit**. This takes you back to the main menu. Be sure to backup if you're not immediately proceeding to Lesson 8.

```
DATE: 27-MAR-87        DATA FLOW - OUTPUT                          PAGE    1
TIME: 15:31            NAME: SUBC: SUBC VIA AD              EXCELERATOR 1.7


TYPE Data_Flow_____        NAME SUBC:_SUBC_VIA_AD_____

    Label SUBSCRIPTION        EXPLODES TO ONE OF:
          VIA                 Record          SUBSCRIPTION_VIA_ADVERTISEMENT__
          ADVERTISE===        Data Model Diagram _____
          MENT_____        ERA Diagram     _____


    Duration Value           _____
    Duration Type            _____

    Description

    Modified By    STUDENT_ANALYST_   Date Modified   870311__   # Changes  8__
    Added By       STUDENT_ANALYST_   Date Added      861123__
    Last Project   COURSE_PROJECT__
    Locked By      _____   Date Locked     8_____   Lock Status  _
```

Figure 7-23

Output Attributes for an Entity

Establish Naming Conventions for Your Next Project

The role of the modern analyst has undergone a shift over the last decade. Increased emphasis has been placed on front-end activities such as requirements analysis or logical design. These shifts have been accompanied by a parallel shift in the primary role of the analyst, from technician to communications specialist. While the ability to communicate technical solutions has not decreased, it proportion to the time spent communicating requirements has dramatically decreased.

To this end, CASE technology is providing better tools for communicating requirements, both business and technical. While EXCELERATOR enforces certain guidelines by standardizing entity types and attribute types, few restrictions are placed on the content of the names and values that you assign. Improper use of names and attributes can be detrimental to the communication process and to the most efficient use of EXCELERATOR's facilities. In other words, EXCELERATOR, like any other tool, is only as good as the people using it. In the remaining sections, we will explore the use of standards for an EXCELERATOR data dictionary.

We used very simple naming conventions for the lessons in Part 2 of the tutorial. By establishing naming conventions, you can further enhance the power of EXCELERATOR's dictionary. Ideally, naming standards should be applied to all projects, the same standards for all projects. However, standards for a single project are better than none. This exercise requires no work on your part. It only suggests some naming standards that may help you with your next EXCELERATOR project. With experience you will undoubtedly develop your own standards and ideas.

Before we begin, let's briefly describe how EXCELERATOR uniquely identifies entity occurrences. Every entity is identified by a three letter abbreviation for its entity type (e.g., *DFD* for data flow diagram, *DAF* for a data flow, *REC* for record, *ELE* for data element, etc.) plus 1-to-32 character ID (which is also known as a *Name* in many EXCELERATOR facilities). Different entities can have the same name so long as they have different IDs. For example, *DAF ORDER* and *REC ORDER* can coexist. It should also be noted that IDs are case sensitive; therefore *DAF ORDER*, *DAF Order*, and *DAF order* are considered three different entities. Finally, do not confuse IDs or Names with Labels. Labels are associated with graphs. They are formatted and frequently more descriptive than IDs or Names.

Let's discuss naming conventions and other possible standards. We'll concentrate on those entities that have been used in the various lessons in this tutorial.

Process Names and Attributes. *Structured Analysis* gives us an ideal ID convention for processes. That convention is based on a simple numbering convention. The context process is always numbered 0. The next level's processes are numbered 1, 2, 3, etc. The children of any of those processes, such as number 2, would be numbered 2.1, 2.2, 2.3, etc. This numbering convention offers several advantages that you may not have considered. For instance, by specifying the name 2* you can generate an output of only those processes in subsystem number 2. Or by specifying 2.1.*, you can specify only a further subset.

You can further refine this ID standard by creating horizontal subsets. For example, by placing the letter P at the end of all primitive processes, you create the option of creating a list or output of all primitive processes. (In the next lesson, you will learn how to generate selection rules for a pointer list to specific entities - and how to generate customized reports from those lists.) Other possible subsets could be defined by character strings (before or after the standard numbers) as follows:

CON	System context (context DFD)
SYS	Major subsystems (system level DFD)
FUN	Major functions of a subsystem
ACT	Major activities of a function
PRM	Primitives (an alternative to P)

Take particular advantage of the process attributes, **Location, Manual or Computer, Duration Type,** and **Duration Value.** They can define alternative groups that can be easily reported. For instance, a report of all computerized primitive processes would be possible, as would a report of all processes performed on a daily or weekly or monthly basis. To make this work, you must decide on standard attribute values.

External Entities Names. We used a very simple EX1, EX2, etc. ID scheme in Part 2. This can be somewhat restrictive when using some of the output options of the dictionary. For example, if you ask the dictionary to list the external entities depicted on a particular DFD, the response would look something like this:

EX2
EX5
EX9
EX10

This doesn't tell you much, unless you look up those entities. For this reason, we recommend that you adopt the following alternative naming scheme. Give the entity a name that abbreviates, approximates, or matches the Label in your DFDs. Maintain alphabetical consistency to make Selector Lists easier to scan. The following examples are offered:

ID	Label
CUST	CUSTOMER
CUST SERV SYS	CUSTOMER SERVICES INFORMATION SYSTEM

By keeping the ID short, you make it easy to retype for purposes of regenerating the label. By keeping the ID close to the label, you make it easier to interpret reports that output only IDs (no labels) and to select IDs off long Selector Lists.

Data Store Names and Attributes.   There are some powerful standardization prospects for data stores. First, now that you understand how to draw DFDs and use the dictionary, you can abandon our oversimplified DF1, DF2, DF3, etc. convention. The reasons are the same as for external entities. Some reports only list IDs. Our IDs wouldn't be very helpful.

For this reason, we recommend that you adopt the following alternative naming scheme. Give the data store a name that abbreviates, approximates, or matches the Label in your DFDs. Maintain alphabetical consistency to make Selector Lists easier to scan. The following examples are offered:

ID	Label
CUSTOMERS	CUSTOMER FILE
PEND ORD	PENDING ORDERS FILE

By keeping the ID short, you make it easy to retype for purposes of regenerating the Label. By keeping the ID close to the label, you make it easier to interpret reports that output only IDs (no labels) and to select IDs off long Selector Lists.

You can also use the attributes to generate interesting subsets for your data store names. For instance, adopt a standard for completing the **Location** attribute:

FC	file cabinet
RD	rolodex
BK	book
BD	binder
DB	computerized data base
FL	computerizes file

The shorter the abbreviation, the lesser the amount of typing and errors. On the other hand, we prefer the shorter descriptive terms for the benefit of the users. Now you can generate a list of all databases currently in use.

The **Manual or Computer** attribute helps you segregate files that are or are not computerized.

Data Flow Names and Attributes.   Data flow IDs don't print on DFDs. They also accumulate in large numbers. Reusability becomes extremely important since IDs are used to check level-to-level balancing (next lesson). Reusability is enhanced through clever, alphabetic abbreviations for labels. This makes IDs and their graphic labels easier to find in Selector Lists.

ID	Label
CUST ORD	CUSTOMER ORDER
VAL CUST ORD	VALIDATED CUSTOMER ORDER
PEND ORD	PENDING ORDERS
ORD COUP W/SUBSC	ORDER COUPON WITH SUBSCRIPTION

Like processes, data flows benefit from responses to the **Duration Value** and **Duration Type** attributes.

We have also found it useful to preface the names of net system input and output data flows with the strings *INP* and *OUT*. This allows us to generate reports on net inputs and outputs using the name ranges *INP** and *OUT**.

Data Entity Names and Attributes. Follow naming conventions similar to those suggested for data stores except that the term *file* should never be used since a data entity does not represent a file.

Data Record Names. As a general rule, data records should always be given identical IDs of the data stores, data flows, or data entities from which they are exploded. There are several advantages. First, it simplifies the standard. Second, you can easily generate reports based on IDs, giving you both the object and record for that object. This naming scheme is easy to implement since recorded are exploded from other objects. When defining the explosion path, just look at the top of the screen for the ID of the entity being exploded.

Data Element Names and Relationships. Data elements are usually numerous; therefore, it is extremely important to use descriptive names. When defining the date of an order, use *ORDER DATE*, not *DATE*. Also, because elements are contained in records, include an abbreviation of the record name at the beginning of the ID. For instance, prefix all elements that are contained in the *CUSTOMER* record with the string *CUST* (e.g., *CUST NUMBER, CUST NAME, CUST BALANCE, CUST CREDIT RATING*, etc.). Consequently, you could easily use the wildcard *CUST** to look at that subset of all elements that defines the *CUSTOMER* record and data store (or entity or flow).

These suggestions are just that -- suggestions! You will likely adopt your own standards for naming and attributes. It is most important to write them down and distribute them to all EXCELERATOR users. That way, everybody can benefit from the standards. EXCELERATOR will reward the use of standards with even greater report flexibility!

Conclusion

Maintaining a project data dictionary is no longer the difficult task that was once perceived. With the **XLDICTIONARY** facility, you can easily predict the effects of changes, ripple many changes through an entire specification document, protect the system from unauthorized changes, and output dictionary contents.

The basic unit of information in the data dictionary is the *entity* (not to be confused with *data entity*). Entities correspond to what data processors call records. Every entity is identified by an *entity type* (which corresponds to what data processors call *record types*). Different entity types are associated with one another by *relationship types* and *classes*. For instance, the entity type *data flow diagram* has a relationship type *contains* with the entity types *process, data store, data flow,* and *external entity*. The combination of entity types and relationship types allow EXCELERATOR to predict and, in some cases, automatically ripple changes through a set of system specifications.

The **XLDICTIONARY** facility is used to inspect entities and relationships, modify entities, and track (audit) all updates to entities. You have learned how to use these capabilities in this lesson.

So far, creating and maintaining system specifications have dominated the lessons in this tutorial. In reality, it has automated what used to be a manual task. However, EXCELERATOR doesn't keep us from making mistakes in our specifications. But EXCELERATOR does provide several useful tools for identifying consistency, completeness, and accuracy errors. That is the subject of the next lesson.

Lesson Eight: Analyzing Documentation Quality Using EXCELERATOR

The Demonstration Scenario

You are rapidly completing your system specification for the record club. Soon, you will turn over the specifications to an application programming team. As a preface to this task, you schedule a meeting with the Assistance Director of Applications Programming. You have used Structured Systems Analysis and Design (SSAD) to develop your specifications. This causes some concern on the part of the Assistant Director, as the following conversation suggests:

> *"This is not my first experience with SSAD. We have noticed some disturbing anomalies in SSAD specifications. First, data flow diagrams and structure charts are nice graphical tools; however, their value is limited by the consistencies between levels. Second, we have noted a marked tendency to find incompleteness in the accompanying dictionary and logic specifications. For example, data elements for which no editing rules have been documented can cause us programming delays. Third, we have also noted numerous inconsistencies in structured specifications. For instance, several different names might exist for a single element. Mind you, we're not criticizing your use of SSAD, only what we perceive to be unavoidable problems with that methodology. Unfortunately, those weaknesses can be a programmer's nightmare! What guarantee do I have that your specifications won't cause delays in the programming phase?"*

The above concerns have been expressed by many managers about other methodologies. How would you respond? Fortunately, CASE products like EXCELERATOR include numerous analytical facilities that can greatly improve the overall quality of all system specifications. The purpose of this lesson is to teach you how to ensure quality in order to produce specification documents that, in all likelihood, are unparalleled in completeness, consistency, and accuracy.

What Will You Learn in this Lesson?

In all but the smallest systems, the quantity of specifications and their interrelationships is substantial, so substantial that the probability of errors is high. These errors include inaccuracies, inconsistencies, and incompleteness. Research has suggested that the cost of fixing these errors increases exponentially over time - that is, an error that would cost $10 to fix in analysis may cost $100 to fix after design and $1000 to fix after implementation. The ripple effects of changes become more complex as the system progresses from a conceptual, paper-based system to a technical, computer-based system. Consequently, any tools that can enhance your ability to catch errors early in analysis and design are economically beneficial to the lifetime costs of the system.

EXCELERATOR provides a number of analytical tools that are immediately available to the analyst. We expect this facet of CASE technology to expand significantly in the next decade. The EXCELERATOR quality checks are provided through a facility called **ANALYSIS**. The facility consists of three sub-facilities:

o   **Graph Analysis**. Graph analysis provides four tools for checking graphs, mostly data flow diagrams.

o   **Report Writer**. This facility allows you to create customized analytical reports off the data dictionary. It requires knowledge of the dictionary's structure and terminology (which you learned in Lesson 7).

o   **Entity List**. Entity list creates ordered lists that are a subset of entities stored in the dictionary. They can be reported directly or used by **Report Writer** to generates customized reports.

These facilities, in conjunction with XLDictionary, assist with the following types of quality checks:

o   Complete specifications

   *   No missing graphs or records. Graphs that have *explosion paths* have been exploded.
   *   No missing dictionary attributes. For example, *edit rules* have been specified for all data elements.

o   Consistent specifications

   *   Consistency between explosion levels of graphs such as data flow diagrams.
   *   No redundant items. For instance, in how many data stores is a data element stored?
   *   Naming consistency between items. Does each data element have one and only one name?

o   Accurate specifications

   *   No obvious graphical errors. For instance, are there any connections which are freestanding from all other objects on the graphs.
   *   Inputs and sufficient to produce outputs.

Many data processing shops have established Quality Assurance staffs and procedures to deal with the growing problem of inconsistent, incomplete, and inaccurate specifications. EXCELERATOR offers you computer-assisted quality assurance. You will learn some of these capabilities in this lesson. Specifically, you will have mastered this lesson when you can:

1.   Generate a report that verifies the mechanical correctness of data flow diagrams.
2.   Generate a report that checks level-to-level balancing and fix any errors identified.
3.   Generate a report that checks for completeness in your specifications.
4.   Generate a report that checks for inconsistency in your specifications.
5.   Generate a pointer list to track dictionary relationships that EXCELERATOR does not track.
6.   Describe how entity lists can greatly enhance analysis and customized analysis reporting.

Before we begin, we should recognize that you have already studied one of EXCELERATOR's analytical facilities, **XLDICTIONARY**. With **XLDICTIONARY**, you can do the following:

o   List, summarize, or inspect dictionary entities by name or wildcard.
o   List occurrences of relationships between entity types that are automatically tracked by EXCELERATOR.

151

These capabilities were covered in Lesson 7. They can be greatly enhanced by intelligent naming conventions such as those that were described at the end of Lesson 7. And they can be further enhanced by the **ANALYSIS** facilities described in this lesson.

Exercise 8.1   Check Data Flow Diagram Mechanics for Validity

This lesson is appropriate only if you completed Lesson 4. In Lesson 4 you drew data flow diagrams. The **ANALYSIS** facility includes an option called **Verification Report**. This report checks for the following simple, mechanical DFD errors:

o   Freestanding objects. Data flow diagrams should contain no freestanding objects. A freestanding object usually indicates incompleteness (missing data flows) or redundancy (duplicate object not deleted) in the DFD.

o   Illegal data flows. Illegal data flows include:

*   Connections between external entities and data stores. Data stores can only be read and updated by processes. This error is corrected by adding the missing processes. Some analysts question this error in instances where a data store is not subject to change and, consequently, outside of the scope of the system (that is, maintained by external entities and used only by the system being designed). However, this situation is properly diagramed by not depicting the external entity; after all, the maintenance of the data store is external to the scope of the project.

*   Connections between two external entities. If both entities are external to the scope of the project, data flows between them are irrelevant to the project.

*   Connections between two data stores. This is a conceptual error. Data can be transferred from one store to another only by way of a process. This error usually indicates that the process is missing. Alternatively, it could indicate that the two data stores are actually a single store.

Step 1.   Before you can demonstrate the report, you need to introduce some of the above errors into your data flow diagrams. Using the **GRAPHICS** facility, introduce the following errors to a Copy of the *SUBSCRIPTION PROCESSING* data flow diagram (see Figure 8-1):

A   **COPY** the *APPLICATION FILE CABINET* data store as a freestanding object, such that it appears twice on the DFD.

B   **CONNECT** the *MEMBER* and *POTENTIAL MEMBER* external entities with a data flow labeled *REFERRAL ORDER FORM*.

C   **CONNECT** the *APPLICATION FILE CABINET* and *PAST MEMBER FILE CABINET* data stores with a data flow labeled *RENEWED APPLICATION*.

Return to the Main Menu.

Step 2.   SELECT ANALYSIS. SELECT **Graph Analysis** (Figure 8-2). This takes you to a Graph Analysis Menu (Figure 8-3). SELECT **Verification Report**. Respond to the name prompt by typing the name of your copied and altered DFD. SELECT **Screen** or **Printer** as your output media.

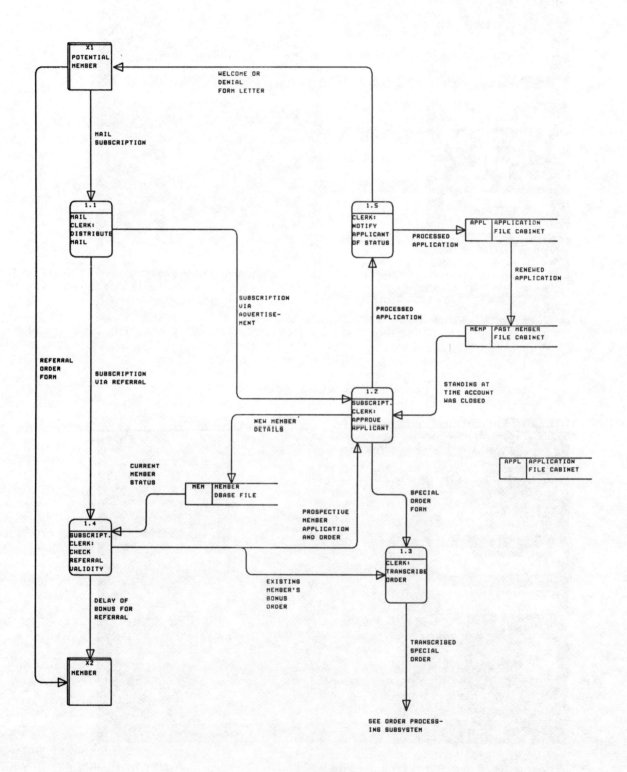

Figure 8-1

Modified Data Flow Diagram

153

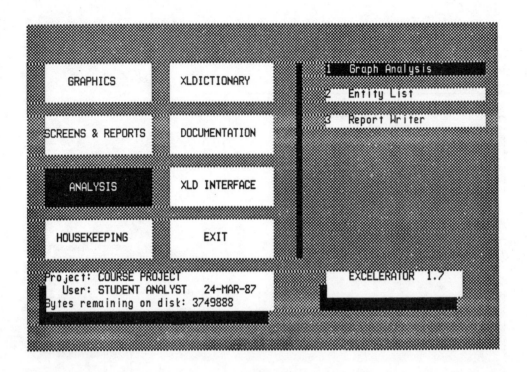

Figure 8-2

ANALYSIS Submenu Screen

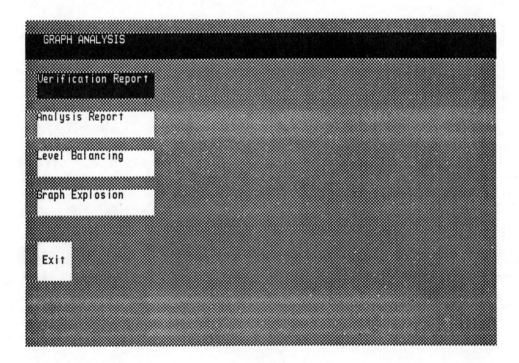

Figure 8-3

Graph Analysis Menu

Step 3.   If you saved the altered DFD, you should get a report similar to Figure 8-4. It is divided into two sections, *FREE STANDING OBJECTS* and *ILLEGAL CONNECTIONS*. The sections are read as follows:

o   The **TYPE** data store (*DAS*) with **ID** *APPL*, **LABEL**ed *APPLICATION FILE CABINET* is a freestanding object.

o   The data flow, *APPL: RENEW APPLIC*, **LABEL**ed *RENEWED APPLICATION*, is illegal because it goes **FROM** a data store (*DAS*) **TO** a data store (*DAS*).

o   The data flow, *ORD: REF ORD FORM* (remember that IDs don't show up on the DFDs), **LABEL**ed *REFERRAL ORDER FORM*, is illegal because it goes **FROM** an external entity (*EXT*) **TO** an external entity (*EXT*).

Exercise 8.2   Check Level-to-Level Balancing of Data Flow Diagrams

Like the last lesson, this lesson is appropriate only if you completed Lesson 4. In Lesson 4 you drew data flow diagrams. The **ANALYSIS** facility includes an option called **Level Balancing**. If you are familiar with the Structured Analysis methodology, you know that balancing describes the consistency between two adjacent explosion levels of a set of data flow diagrams. Levels of DFDs are balanced if, for any DFD at a higher level, the data used by or produced by that DFD is completely accounted for in any subsequent levels of decomposition. This ensures that you haven't lost any details in the leveled set.

In pure *Structured Analysis*, data flows should balance all the way up to the context diagram. In practice, this would result in excessively complex, cluttered context, system, and middle-level DFDs. Consequently, we suggest that data flows balance down to lower levels (to ensure consistency), but not necessarily upward (to higher levels). Upper level diagrams show only the highest volume or most important data flows (deferring trivial or less common flows to the lower levels).

This is roughly how EXCELERATOR's balancing report works. It checks any specified DFD against all the DFDs to which its processes directly explode. Then it reports errors to you. Let's execute the **Level Balancing** report. If you completed the DFDs in Lesson 4 exactly as we did, then there are some level balancing problems that will allow you to demonstrate this important feature.

Step 1.   SELECT ANALYSIS. SELECT **Graph Analysis**. SELECT **Level Balancing**. You will be prompted for a name. Type *RECORD CLUB SYSTEM DIAGRAM* or select that graph from a Selector List. EXCELERATOR will prompt you for the **# of levels** (Figure 8-5). EXCELERATOR will check up to nine levels below the graph you have named. It will divide the analysis report into sections, each section corresponding to one DFD and all of the DFDs to which its processes explode. We want you to retain the default value **1**. Press the Enter key. EXCELERATOR now prompts you for **Screen, Printer,** or **File** as the output medium. SELECT **Printer** or **Screen**. Press the Enter key to execute the report.

Step 2.   You should get a report very similar to the one shown in Figure 8-6 (provided you did not correct our balancing errors in Lesson 4). The top of the page is header information. Most of it is self-explanatory; however, it should be noted that the Level Number for the graph that you initially selected is always *0*. The errors are interpreted as follows (you should have printouts of your DFDs or our DFDs from Lesson 4 as reference):

o   First notice that the report has no messages related to the process whose ID (name) is *1* and whose label is *SUBSCRIPTION SUBSYSTEM*. That's because there were no balancing errors detected for this process.

o   For the process whose ID (name) is *2* and whose label is *PROMOTION SUBSYSTEM*:

*   The **OUT**put data flow *DATED ORDER* (which appeared on the parent DFD) does not appear on the child DFD (the DFD to which process 2 explodes). If you check the child diagram, called *PROMOTION PROCESSING*, you will find

	FREE STANDING OBJECTS:				
TYPE	ID	LABEL			
DAS	APPL	APPLICATION FILE CABINET			

	ILLEGAL CONNECTIONS:				
ID		LABEL		FROM	TO
APPL: RENEW APPLIC ORD: REF ORD FORM		RENEWED APPLICATION REFERRAL ORDER FORM		DAS EXT	DAS EXT

Figure 8-4

Graph Analysis Report

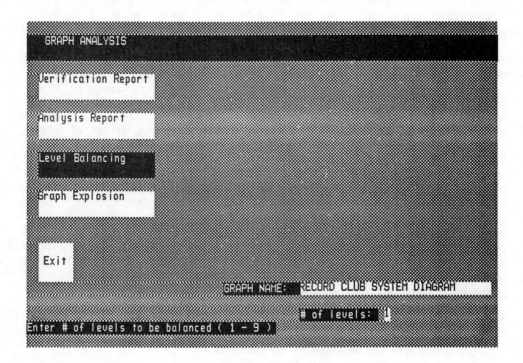

Figure 8-5

Request for Level Balancing Report

156

```
DATE: 24-MAR-87                         DATA FLOW DIAGRAM  LEVEL BALANCING                   PAGE    001
TIME: 12:04                                                                                  EXCELERATOR
GRAPH NAME: RECORD CLUB SYSTEM DIAGRAM                   GRAPH FILE: E3N2ZY0.DFD

Level Number:      0
Parent Graph Name:   RECORD CLUB SYSTEM DIAGRAM

Entity        Prnt
 Type    Dir  Chld
------- ----  ----  ---------------Name (ID) or Error Message-------------- -----------------------------Label-----------------------
                                                                         PROMOTION SUBSYSTEM
Process        P    2
Graph          C    PROMOTION PROCESSING
Flow     OUT   P    ORD: DATED ORD                                       DATED      ORDER
                    ***Error  -  Parent Data Flow does not match any Child Flows  (refers to above item(s))
                                                                         RECORD     PROMOTIONS
Flow      IN   P    PROMO: REC PROMO
                    ***Error  -  Parent Data Flow does not match any Child Flows  (refers to above item(s))

                                                                         ORDER      PROCESSINGSUBSYSTEM
Process        P    3
Graph          C    ORDER PROCESSING
Flow      IN   P    MEM: CUR MEM STATUS                                  CURRENT    MEMBER     STATUS
                    ***Error  -  Parent Data Flow does not match any Child Flows  (refers to above item(s))
                                                                         ORDERS
Flow      IN   P    ORD: ORDERS
                    ***Error  -  Parent Data Flow does not match any Child Flows  (refers to above item(s))

                         *****END OF REPORT*****
```

Figure 8-6

Level Balancing Report for Data Flow Diagram

that we named that data flow *COPY OF ORDER TICKET*. Again, it is common to give more descriptive names as we draw more detailed diagrams. Still, we should have used consistent names between the levels.

This error can also be easily corrected. The data flow IDs for both diagrams must match. Again, this correction can only be accomplished through the **GRAPHICS** facility. Fix the error at this time.

* The **OUT**put data flow *RECORD PROMOTIONS* (which appears on the parent DFD) does not appear on the child DFD. The problem is the same. On the child DFD, the flow was called *RECORDS OF MONTH OR SPECIAL PROMOTIONS*. This technique to fix the error is the same. Fix the error at this time.

o For the process whose ID (name) is *3* and whose label is *ORDER PROCESSING SUBSYSTEM*:

* The **IN**put data flow *CURRENT MEMBER STATUS* does not appear on the child DFD (the one exploded from process *3* on the parent DFD). What happened is that, when we drew the child DFD, we realized that we needed more than the status to process the order. Indeed, we needed various *MEMBER DETAILS* (e.g., names, address, etc.) Consequently, we renamed the flow on the child diagram and forgot to do the same on the parent diagram. The fix has already been described. Fix the error at this time.

* Now for a different type of error. The **IN**put data flow *ORDERS* on the parent DFD does not appear on the child DFD (the one exploded from process *3* on the parent). This is not really an error! Analysts frequently consolidate flows at the upper levels to keep those DFDs readable. That is precisely what we did. At the upper level we used the term *ORDERS*. We did not feel it appropriate to differentiate between types of orders until we were drawing the DFD whose processes dealt with specific types of orders.

You could simply ignore the error message; however, you can easily eliminate the error message by describing the equivalence to the data dictionary. We describe the technique in the next few steps.

Step 3.   **Exit** from the **ANALYSIS** facility to return to the Main Menu. **SELECT GRAPHICS.** SELECT **Data Flow Diagram** from the Graphics Menu. SELECT **Modify** from the Action Keypad. When prompted for a name type *RECORD CLUB SYSTEM DIAGRAM* and press the Enter key (or that name from the Selector List).

Step 4.   You should be looking at the DFD. SELECT **DESCRIBE** from the Graphics Commands Menu. SELECT the *ORDERS* data flow from the diagram. Press the Enter key when prompted for the ID and select *ORD: ORDERS* from that list (this should all be old hat to you). You should now be at the familiar Data Flow Description Screen (Figure 8-7). What we want to do is to define an explosion path for *ORDERS*. SELECT the first position in the **Record** line of the **Explodes To One Of** block. Type *ORDERS*. Press the F3 function key to save your changes. This takes you back to the DFD.

Step 5.   SELECT **EXPLODE.** SELECT the *ORDERS* data flow from the graph. This takes you to a Record Description Screen for *ORDERS*. Complete the screen as shown in Figure 8-8. Notice that we have used names that correspond to the order types. Also note that we gave each name the type *R* for record. There is no need to describe the specific contents of these subrecords to correct our balancing error.

Press the F3 function key to save the changes (and return to the graph).

Step 6.   SELECT **EXPLODE.** SELECT Process 3 from the graph. This takes you to the child DFD where we first differentiated between the two types of *ORDERS*. Each of these data flows (labeled *MAILED IN ORDER COUPON IN RESPONSE TO PROMOTION* and *SPECIAL ORDER AND/OR BONUS COUPONS*) needs to be described down to the record level.

SELECT **DESCRIBE** and the data flow *MAILED IN ORDER COUPON IN RESPONSE TO PROMOTION* from the diagram. Press the Enter key. This takes you to the Data Flow Description Screen. Describe the data flow as exploding to a record of a similar name. Press the F3 function key to save the changes and return to the diagram. Now SELECT **EXPLODE** for the same data flow. This takes you to a Record Description Screen. Complete it as shown in Figure 8-9. The sub-record name must exactly match the name you typed during Step 5. Repeat this procedure for the *SPECIAL ORDER AND/OR BONUS COUPONS* data flow.

Time out! Did you surmise what we were doing in Steps 3 through 6? It really is relatively simple. First you exploded the *ORDERS* parent data flow into two records that corresponded to the two child data flows. Then you exploded each of the child data flows into records. The balancing analysis will realize that the flows are logically equivalent and, therefore, not print an error message.

This is an appropriate time to explain how EXCELERATOR checks for balancing errors. Level balancing looks at two levels at a time: a parent diagram and all of the child diagrams to which it explodes. For any given process on the parent diagram, it checks the data flows to/from that process against the net data flows on the corresponding child diagram. A net data flow is defined as one of the following:

o   The data flow to/from an object has the same ID (name) in both the parent and child diagrams.
o   The data flow is actually an *interface* that connects an object to an object outside the diagram (on another diagram). You learned about interfaces in Lesson 4.

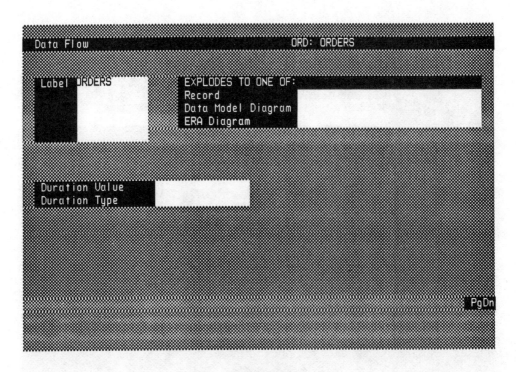

Figure 8-7

Data Flow Description Screen

Record	ORDERS

Alternate Name
Definition

Name of Element or Record	Occ	Seq	Type
MAILED IN ORDER COUPON	1	0	
SPECIAL ORDER / BONUS COUPONS	1	0	
	1	0	
	1	0	
	1	0	
	1	0	
	1	0	
	1	0	
	1	0	
	1	0	
	1	0	
	1	0	
	1	0	

PgDn

Figure 8-8

Record Description Screen for a Data Flow

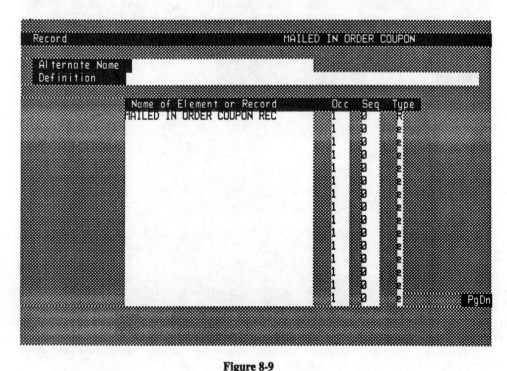

Figure 8-9

Modified Record Description Screen

EXCELERATOR checks the above flows as follows:

1. First, the direction of the flows is checked to make sure they are the same on both levels.
2. Second, the source of the data flows is checked to make sure they match at both levels. This check does not apply to *interface* data flows. If the source of the flow is a data store on the parent diagram and an external entity on the child diagram, an error message would result.
3. Third, EXCELERATOR checks the IDs of the flows on the parent and child diagrams. For matching data flows with matching IDs, those flows are considered balanced. For IDs that don't match, the next step is performed.
4. Fourth, flows that didn't match by ID are examined to see if they explode to the same records. This check works like a mathematical *union*. For example, if parent data flow X consists of records A, B, and C, and the child data flow Y1 explodes to record A, the child data flow Y2 explodes to record B and the child data flow Y3 explodes to Z, then the data flows are considered balanced. That is precisely what we did in steps 3-6.[1] If this test fails, the process moves to the next step.
5. Finally, any data flows which still don't balance are checked at the decomposition level. This means EXCELERATOR compares the contents down to the element level. The mathematical union concept is still applied. Of course this step assumes that you have defined data flows down to the element level. Any data flows that still don't balance after this test are reported as errors.

An even more detailed explanation of level balancing can be found in InTech's *EXCELERATOR Reference Manual*. Let's return to the exercise ...

Step 7.   Since you've corrected all of the balancing errors in the parent DFD, you can now repeat Step 1. The new balancing report should only contain the header data (identifying the graph) and the message *****END OF REPORT*****. This signifies that the two levels balance.

Step 8.   OPTIONAL. Run the balancing reports for all of your non-primitive data flow diagrams. If you followed the DFDs we presented in Lesson 4, you should find a few errors. Correct the errors to produce a completely balanced set of DFDs. Re-run the balancing reports to prove correctness.

That completes our exercises on graph analysis. It should be noted that there are other graph analysis reports that you may find useful. They are fully documented in InTech's manuals. EXCELERATOR also includes dictionary analysis facilities. They are covered in the next two exercises.

Exercise 8.3   Create Your Own Customized Analysis Report

The **Report Writer** facility lets you create customized and reusable analysis reports from dictionary entities. You can choose to base your report on one of EXCELERATOR's predefined entity types, relationship classes, or relationships types. You can specify your own selection rules for the entities to be included in your report. Report definitions can be saved for later use. They can also be transported (through the **XLD INTERFACE** facility) to other projects. Over time, an entire library of reports can be defined and shared by EXCELERATOR users.

We could actually devote an entire lesson to a facility with the potential of **Report Writer**. However, our purpose is to give you some flavor of the power of the facility, not to overwhelm you. We have divided the coverage of **Report Writer** into separate exercises, one to analyze the completeness of specifications, and another to more carefully present your options for future analysis reports.

Consider the following scenario. A typical completeness check might include a report that ensures that all data elements have **Edit Rules**. This is a reasonable standard. Even for fields like *CUSTOMER NAME*, it is reasonable to expect a statement that there are no value limitations. Otherwise, the programmer doesn't know whether you forgot or whether there are no limitations on values. Before using **Report Writer**, you should ensure that the report cannot be generated from the **XLDICTIONARY** facility. This report clearly could not (unless you want to wade through the complete output descriptions of *all* data elements). But the necessary data is in the dictionary. That is the prerequisite for use of the customized **Report Writer**.

Step 1.   SELECT **ANALYSIS**. SELECT **Report Writer**. An Action Keypad appears. Once you have defined (or imported) some custom reports, you can use the **Modify** or **Execute** actions. However, since this is our first report SELECT **Add**. When you are prompted for a name, type *INCOMPLETE DATA EDITING*. It is a good idea to give all customized reports both descriptive **Names**. This way, others can use the customized report. Press the Enter key.

This brings up a Selector List of **XLD Types**. SELECT the *ELE* (which stands for *data element*).

Once an **XLD Type Code** is established for a report, it cannot be modified.

Step 2.   This takes you to a Report Definition Screen (Figure 8-10). First, you need to define a scope for the entities to be included in your report. Our report can be answered directly via **Report Writer**; therefore, SELECT **Select** from the Report Definition Screen.

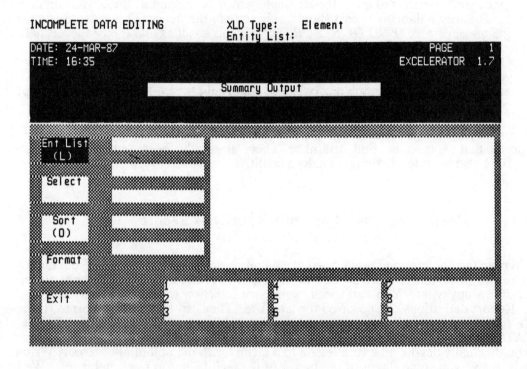

INCOMPLETE DATA EDITING   XLD Type: Element
                                 Entity List:

DATE: 24-MAR-87   PAGE   1
TIME: 16:35   EXCELERATOR 1.7

                        Summary Output

Ent List
  (L)

Select

Sort
 (O)

Format

Exit   1   4   7
                2   5   8
                3   6   9

Figure 8-10

Report Definition Screen

Step 3.   A Selector List of attributes appears. Now we must define the selection rules. Notice that every attribute of ELE appears in the list. The first attribute, though unnecessary, is included for practice. We want to examine *all of* the data elements (the default). SELECT **Name.** A window appears for the selection rules for *Name.* Type "*" (the quote marks must be included for all non-numeric attributes) in the first rule bar (Figure 8-11). This will select all the data elements. Now press the F3 key to save this selection attribute and associated rule.

The second selection rule is based on the attribute *edit rules*. SELECT **Edit Rules** from the Selector List. Once again, the selection rules window appears. In the first rule line, type *NOT "*" OPTIONAL* and press the F3 key. This tells EXCELERATOR to ignore any data element that has any value in the *edit rule* attribute. All remaining elements have no edit rules at this time.

That represents all the rules for this report. You could have defined up to five attributes and rules. Signify completion by SELECTing the attribute **...Complete...** . EXCELERATOR will ask you if it is OK to proceed. Type *Y* for *yes.* A summary of the attributes you selected and the rules now appear.

Step 4.   Normally, the selected data elements would be sorted alphabetically by name (ID). You can, however, sort them on the basis of any attribute that you wish. For demonstration purposes, let's sort the elements on the basis of date created. For elements created on the same date, we'll sort them alphabetically.

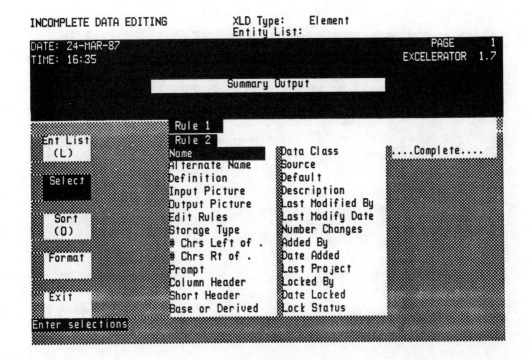

INCOMPLETE DATA EDITING   XLD Type: Element
                                           Entity List:

DATE: 24-MAR-87   PAGE   1
TIME: 16:35   EXCELERATOR 1.7

                              Summary Output

         Rule 1
Ent List Rule 2
 (L)   Name   Data Class   ...Complete....
         Alternate Name Source
 Select Definition   Default
         Input Picture   Description
         Output Picture Last Modified By
 Sort   Edit Rules   Last Modify Date
 (O)   Storage Type   Number Changes
         # Chrs Left of . Added By
 Format # Chrs Rt of . Date Added
         Prompt   Last Project
         Column Header   Locked By
 Exit   Short Header   Date Locked
         Base or Derived Lock Status
Enter selections

Figure 8-11

Rule Bars for Selection of Entities for a Customized Analysis Report

SELECT **Sort**. A Selector List of attributes is displayed. SELECT **Date Added**. Type the number *1* to indicate that it is a first-level sort key. You could repeat this paragraph for the attribute *Name*; however, since EXCELERATOR defaults to *Name*, it would be redundant (and, incidentally, consume unnecessary time to produce an index that already exists.

Step 5.   Next, you need to specify the output options for your customized report. SELECT **Format**. A submenu appears (Figure 8-12). The submenu defines format options as follows:

o   **User-Defined.** This option allows you to select the attributes to be reported as well as the format of the report.
o   **Output.** This predefined report format outputs all attributes for the selected entity names.
o   **Summary Output.** This predefined report format outputs a list of the selected entity names and two key attributes (that vary from one entity type to another).
o   **List.** This predefined report format displays (but cannot print) a list of the selected entity names and one key attribute (that varies from one entity-type to another)
o   **Audit Output.** This predefined report format outputs only the audit attributes for the selected entities.

The last four options are identical to the output options that you learned in the **XLDICTIONARY** facility. For ultimate flexibility, let's SELECT 1 for **User-Defined**.

Step 6.   The User-Defined Menu (Figure 8-13) appears. The top portion of the screen is used to define the page layout of your report (up to 132 columns). The rest of the screen is used to select attributes that you want to appear on the report.

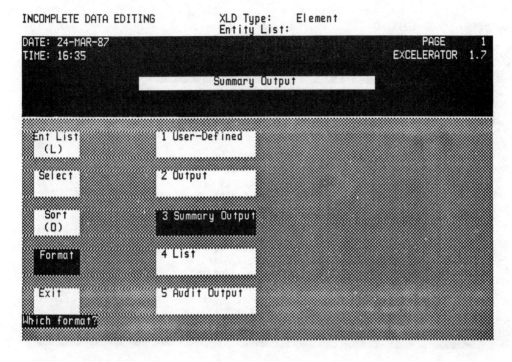

Figure 8-12

Customized Analysis Report Format Options

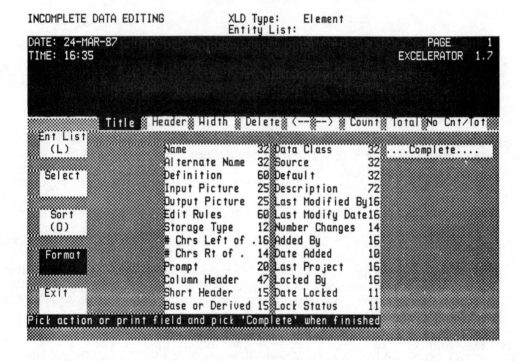

Figure 8-13

User-Defined Report Format Menu

SELECT **Title** by pressing the Enter key (from the horizontal format menu). Type a title and press the F3 function key.

The purpose of our report is to identify those elements which have no specified *edit rules*. We might want such a report to specify attributes that could help define *edit rules*. These might include the following: *name, storage type, base or derived, characters left of decimal, characters right of decimal,* and *input pic.*

From the Selector List, SELECT the first of these attributes, *Name.* Using Figure 8-14 as a guideline, SELECT the starting location for the attribute. A column header will be added by EXCELERATOR. Repeat this process for the remaining attributes, preferably in the sequence presented above.

When you've finished adding all desired attributes to your report, SELECT **...Complete....** This takes you back to the Report Definition Screen.

Step 7. SELECT **Exit.** SELECT **Save.** You should now return to the Report Writer Action Keypad.

Step 8. To print your report SELECT **Execute** from the Action Keypad. Respond as directed. Since you saved the report definition, it can be re-**Executed** at a future time. Experienced EXCELERATOR users will establish a procedure for cataloging and sharing their analytical reports. Any base customized report can be **Modify**(ied) to your own specific needs. You can change selection rules, sorting criteria, and formatting to your own tastes.

```
DATE: 24-MAR-87          REPORT OF ELEMENTS CONTAINING NO EDIT              PAGE      1
TIME: 19:30              RULES                                             EXCELERATOR 1.7

Name                             Storage Type  Base or Derived  # Chrs Left of .  # Chrs Rt of .  Input Picture

NAME                             C             B                10                0               X(10)
PERCENT ORDER RESPONSES          C                              2                 0               99
STREET                           C                              15                0               X(15)
ORDER CREDITS                    C             D                2                 0               99
ORDER DATE                       D             B                8                 0               MMDDYY
PERCENT NO ORDER RESPONSES       C                              2                 0               99
TOTAL ORDER RESPONSES RECEIVED   C                              4                 0               9999
TOTAL REJECTING MONTHLY TITLE    C                              3                 0               999
ZIPCODE                          C                              9                 0               X(9)
TITLE                            C                              20                0               X(20)
TOTAL ACCEPTING MONTHLY TITLE    C                              3                 0               999
TOTAL ACCEPTING MONTLY TITLE     C             D                3                 0               999
CREDIT                           C             B                1                 0               9
CREDITS EARNED                   C             B                2                 0               99
CITY                             C                              15                0               X(15)
AMOUNT                           C             B                5                 0               99999
CATALOG NUMBER                   C                              5                 0               99999
DATE                             D                              8                 0               MMDDYY
NUMBER OF ORDER RESPONSES        C                              3                 0               999
NUMBER OF RESPONSES NOT RECEIVED C                              3                 0               999
  ORDER RESPONSES                C                              6                 0               999999
  TE ENROLLED                    D             B                6                 0               999999
EMBER NAME                       C             B                30                0               X(30)
```

Figure 8-14

Customized Report Format

165

The Power of Report Writer

Now that you understand the basic procedure for defining customized analysis reports, we want to present a more detailed explanation of the Report Definition Screen options. This is not an exercise; however, it is an important part of learning how to put the potential of report writer to work for you. The presentation is organized according to the two key steps in customized analysis report definition:

1.   Define selection attributes, selection rules, and sorting rules.
2.   Define report format.

Define Selection Attributes, Selection Rules, and Sorting Rules. Selection attributes and selection rules always occur in matching pairs. Selection Attributes are usually triggered by SELECTing **Select** from the Report Definition Menu. Alternatively, selection can be based on entity lists which are defined in the next exercise.

When you SELECT an attribute from the Selector List, a two-bar Selection Rule window appears. You may place one or more selection rules in either bar. The following rules must be followed:

o   If the rule value is an alphanumeric string, it must be contained in double quotes (e.g. "REC"). Separate multiple values by commas.

o   If the rule value is a number, do not use quotes. If the value is a set of finite numbers, separate numbers by commas.

o   If the rule value is a range of numbers, use one of:

        <   less tan
        >   greater than
        THRU   through

o   You may use wildcards. Wildcards are more flexible in **Report Writer** than in the other EXCELERATOR facilities. Recall that, in most facilities, the wildcards could only occur at the end of a valid string (for instance, *MEMBER** would locate entities whose names begin with the string *MEMBER*). What if you want to retrieve entities that have the *MEMBER* string in their name, but not necessarily at the beginning. Also, suppose that you have used various combinations of upper and lower case. The following rule for the *Name* attribute would retrieve most, if not all, occurrences:

        "*MEMBER*","*member*","*Member*"

o   You can use the following qualifiers:

        NOT   to exclude values as specified above
        OPT   to include blank entries
        NOT OPT   to target blank entries

There are two selection rule bars. If you specify rules in both, only entities which fulfill *both* sets of rules will be included in the report. The above options will accommodate most selection needs. More complex needs must be handled through entity list.

Once entity selection rules have been specified, you may consider whether or not you want sort criteria. You can specify up to nine sort levels but the more the sort levels, the longer the processing time and the more disk storage required. Name, an alphabetically sorted attribute, is the default if no sort criteria is specified. Normally, this is more than adequate. Name is also automatically the last sort level.

Define Report Format. As noted earlier, the predefined report formats simply use the formats of the Action Keypad output options in the **XLDICTIONARY** facility. For user-defined reports, you have the following options:

o   **Title.** You used this command to add a title to your first report.
o   **Header.** This command allows you to modify headers for columns.
o   **Width.** Use this command to adjust the width of a column (field).

o   **Delete.** This command deletes a column (field) from the report.
o   **<-and->.** Use this command to scroll the screen left and right.
o   **Count.** This command counts the blank entries for any given column.
o   **Total.** This command totals values of numeric (only) columns.
o   **No Cnt/Tot.** Remove either count or total options for a column.

With selection, sorting, and formatting, you can create some very useful and powerful analytical reports. The possibilities are limited only by your imagination.

Exercise 8.4   Create an Entity List for Customized Analysis Reports

What if you want to track relationships that EXCELERATOR does not normally track? Or what if you want to generate a customized analysis report that is to be based on a set of entities from more than one, but not all, entity type? (Recall that **Report Writer** only works with one entity type or all entity types.) EXCELERATOR provides a facility, **Entity List**, that helps you address these unique situations.

Recall the following fundamental definitions (from Lesson 7):

o   *Entity.* The standard unit of information recorded in your dictionary. Each entity has a unique name.
o   *Entity Type.* A group of entities that have the same attributes (for instance, ELEment. Each entity type has a three letter identifier).

Essentially, an *entity list* is an index that selects dictionary entities based on the entity types, combinations of entity types, entity attributes, and selection rules for attributes, all of which you define. The index is used by EXCELERATOR to retrieve entity occurrences pointed to by the list. Actual entity lists can be saved, but more importantly, the definition rules that generate entity lists can also be saved (as its own XLDICTIONARY entity-type and entities). This means that a new entity list can be generated for an evolving set of specifications.

What can you do with entity lists? First, the **Entity List** facility contains a complete set of output actions very similar to those provided by **XLDICTIONARY**. Consequently, given an actual entity list, you can inspect the entities, list the entities or relationships, and output the entities (full descriptions, summaries, audit attributes only, etc.). However, the real power of entity lists lies in your ability to generate customized analytical reports about the entities pointed to by the entity list. This facility is very similar to the **Report Writer** facility you just learned.

When should you use entity lists? Entity lists should be used if there are subsets of entities that you frequently use (e.g. by category, by selection rules, or by relationships, or some combination). The entity list can be named, quickly regenerated to bring it up to date, and used in conjunction with **report Writer's** customized report definitions (recall that **Report Writer** contained an option to select entities off an **Entity List.**). When should you *not* use entity list. Entity list is inappropriate for reports and queries that can generated off **XLDICTIONARY's** standard output facilities.

Let's do a very simple entity list. The purpose is to demonstrate how to create the entity list and then output the resulting entities. We'll generate a user-defined report since, if you understand that option, all the other report options are easy (because they are predefined). A subsequent section to this exercise will describe the power of entity lists in greater detail.

Initially, we need a scenario. We want to know all data flows *and* data entities whose records contain the data element *MEMBER NUMBER.* There is no direct EXCELERATOR relationship between data elements, data entities, and data flows. However, the following logical procedure would answer our query:

a.   Generate an entity list that points to all data records *(REC)* that the data element *(ELE) MEMBER NUMBER* is *Contained-In.*
b.   From the first entity list, create a second entity list that contains those data records *(REC)* that *Explode-from* data flows *(DAF).* Only those records on the first list that explode from data flows will be included on the second list.

167

c.   Again from the first entity list, create a third entity list that contains those data records *(REC)* that *Explode-from* data entities *(DAE)*. Only those data records from the first list that explode from data entities will be included on the third list.

d.   Create a final entity list that merges the contents of second and third entity lists. This answers our query.

Note that the above procedure would exclude all records that explode from data stores. Otherwise, we could have answered our query from the **XLDICTIONARY** facility using the *ELE Contain-In ANY* relationship.

The above procedure is exactly how you would use **Entity List**. What would be the value of going to this much trouble? Suppose you need to do this for several data elements, several times during the project. Just plug in the value of a new data element to generate a new list of data flows and data entities! Here's how to generate the list.

Step 1.   From the Analysis Menu, SELECT **Entity List**. SELECT **Add** from the Action Keypad.

Step 2.   From the Entity List Add Option Screen (Figure 8-15), SELECT **XLD Selection**. This tells EXCELERATOR that you want to define some selection rules.

At the bottom of the screen, you will be prompted for **Name** and **XLD Type**. These terms carry exactly the same meaning as they did for **Report Writer**. **Name** should be used to give a meaningful name to the entity list. For **Name** type *RECS CONTAINING MEMBER NUMBER*. Press the Enter key. A Selector List of valid **XLD Types** is displayed. SELECT **ELE-Contained In-REC** (on page 2 of the Selector List). Press the Enter key.

Step 3.   This brings up the XLD Selection Criteria Screen (Figure 8-16). It is very similar to those screens you used in Exercise 8.3, the **Report Writer** exercise. SELECT the attribute **Name**. Type *VALUE IS "MEMBER NAME."* Remember to use double quotes for all non-numeric values. Wildcards, as always, can be used. To save **Rule 1**, press the F3 function key.

We have no other selection criteria for this list; therefore, SELECT **....Complete....** . EXCELERATOR will asks, *OK to proceed?* Type *Y* for *yes*. EXCELERATOR will automatically create the entity list. You are automatically returned to the Action Keypad.

To confirm that the list was created, SELECT **Inspect**. When prompted for a name, press the Enter key. SELECT your entity list from the Selector List. The contents of the entity list will be displayed (they will vary depending on which lessons you completed). Press any non-arrow key to return to the Action Keypad.

It should be pointed out that EXCELERATOR saves the actual entity list, not the rules used to generate that list.

Step 4.   Now we want to create the *DAF Explodes-To REC* and *DAE Explodes-To REC* entity lists. SELECT **Add**. SELECT **XLD Selection from List**. This option is used to further subset the list created in the last step. For **Name**, type *DAF EXPLODES TO REC WITH MEMBER NO.* Press Enter. For **XLD Type**, type *REC Explodes-From DAF*. This is an example of using a relationship as the **XLD Type** subject of our list operation. Press Enter. From the XLD Selection Criteria Screen, SELECT **....Complete....** because we have no additional selection criteria. Once again, EXCELERATOR asks, *OK to proceed?* Type *Y* for *yes*. EXCELERATOR automatically produces the new entity list. The new list contains the names of data flows that contain the desired element - we are half way there! You may **Inspect** the list if so desired.

Now we want to create the DAE Explodes-To REC entity list. We'll use the same procedure we did for the second list. That is, we'll break it out of the first list. SELECT **Add**. SELECT **XLD Selection from List**. For **Name**, type *DAE EXPLODES TO REC WITH MEMBER NO.* Press Enter. For **XLD Type**, type *REC Explodes-From DAE*. Press Enter. From the XLD Selection Criteria Screen, SELECT **....Complete....** because, once again, we have no additional

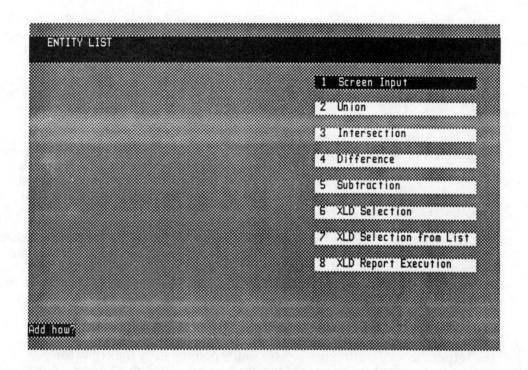

Figure 8-15

Entity List Add Option Screen

ELE CONTAINED IN REC 401   XLD Type: Element
                                Entity List:

Name	Data Class	....Complete....
Alternate Name	Source	
Definition	Default	
Input Picture	Description	
Output Picture	Last Modified By	
Edit Rules	Last Modify Date	
Storage Type	Number Changes	
# Chrs Left of .	Added By	
# Chrs Rt of .	Date Added	
Prompt	Last Project	
Column Header	Locked By	
Short Header	Date Locked	
Base or Derived	Lock Status	

Pick selection field and pick 'Complete' when finished

Figure 8-16

XLD Selection Criteria Screen

169

selection criteria. Once again, EXCELERATOR asks, *OK to proceed?* Type *Y* for *yes*. EXCELERATOR automatically produces the new entity list. The new list contains the names of data entities that contain the desired element. Again, you may **Inspect** the list if you wish.

Step 5.   We now have separate lists for the element contain in data flow and data store. Can they be combined. Yes! While the two lists, as they are may be sufficient, let's combine them to demonstrate another feature of **Entity List.**

SELECT **Add.** This time, SELECT **Union.** This option allows you to merge the contents of two entity lists, eliminating duplicate entries. The resulting list will be our final entity list. For **Name,** type *ELE MEMBER NO IN DAF OR DAE.* The two prompts below the name of our final entity list refer to the entity lists that you wish to merge. For **Using List,** type *DAF EXPLODES TO REC WITH MEMBER NO* (the second entity list). For **And List,** type *DAE EXPLODES TO REC WITH MEMBER NO* (the third entity list). EXCELERATOR will prompt, *OK to proceed?* Type *Y* for *yes*. EXCELERATOR will automatically construct the lists from union of the two specified lists. You will also return to the Action Keypad.

Step 6.   All we have is an index into entities. You can execute any of the following output options (all of which should be familiar to you by now):

o   **List** the entities pointed to by the entity list.
o   **Inspect** the contents of the entity list itself.
o   **Output** the all the attributes of each entity pointed to by the entity list.
o   **Summary output** the name and two predefined attributes of every entity pointed to by the entity list.

All of the above options are available through the Action Keypad. They behave exactly as they do in every other facility you've studied. At this time, you may experiment with these output options if you so desire.

You can create customized reports (based on selected attributes and column locations) from entity lists. You learned how to use **Report Writer** in the last exercise. The only difference is that, when using entity lists, instead of SELECTing **Select,** you would SELECT **Ent List** and provide the name of the entity list.

The Power of Entity Lists

Entity lists are one of EXCELERATOR's most powerful concepts. Now that you understand how to create them, you should learn a little more about their power. This description doesn't require that you use EXCELERATOR since it would take an appreciably larger and more complex project dictionary to demonstrate all of the capabilities of entity lists. We'll divide our discussion into subsections dealing with the natural activities of creating, modifying, and deleting entity lists.

Creating Entity Lists.   When you SELECTed **Add,** you probably noticed that there are eight options for creating new entity lists. Each option creates an entirely new entity lists. Each list must be stored on disk. The eight options are described below:

o   **Screen Input.** In this option, you directly type in the entity types and names of the specific entities that you want to include in your list. This should only be used for small, finite lists. It can also be used for lists pointing to items not currently stored in the dictionary (in other words, they cannot be selected).

o   **Union.** The first of four set operations, you used this option in the last exercise. Given two previously built entity lists, this option creates a third entity list that contains all the entries contained in both original lists (but with no duplicates). It is a terrific way to join lists for more than one entity type when you don't want to use the *Any* entity type option.

o **Intersection.** The second of four set operations, this option also builds a new entity list from two existing entity lists. The difference is that the resulting list contains only entities that appear in both of the original lists.

o **Difference.** The third of four set operations, this option also builds a new entity list from two existing entity lists. The resulting entity lists contains entries that appeared in one of the original lists, but not both of the original lists. Thus, it is the inverse of intersection. This option is the best way to build an entity list of all entities that were changed over a specific time period. By regularly generating an output of this list, you could create a more detailed audit trail than is provided by EXCELERATOR.

o **Subtraction.** The last of the four set operations, this option produces a new entity list that contains the entries that are found only in the first list, but not the second.

o **XLD Selection.** At this time, you are most familiar with this option since you used in Exercise 8.4 and because it is very similar in operation to the selection techniques you learned in the **Report Writer** facility.

o **XLD Selection from List.** As you learned in the last exercise, this facility is similar to the **XLD Selection** option; however, it starts from an existing entity list.

o **XLD Report Execution.** This option creates a new option using an existing report definition from **Report Writer**. The predefined report must have selection criteria that meet the requirements of your entity list.

Modifying Entity Lists. It is very important that you understand what happens when you **Modify** an entity list. Unlike **Add**, **Modify** creates a new version of an entity list. This overwrites the previous version of that entity list. It is important to realize that EXCELERATOR does not save the rules used ti generate the original list. When you SELECT **Modify**, you use one of the same eight options that you used to create the original list. In essence, you really recreate the list.

Sometimes, you SELECT **Modify** only because you need to regenerate a frequently used list to reflect the addition and deletions of entities. Otherwise, you regenerate the list based on new rules or set operations. Finally, **Modify** is the command used to sort entity lists.

Deleting Entity Lists. Entity lists can be very long. If they are frequently generated, and there are numerous benefits to their frequent use, they can consume considerable disk storage. Therefore, it is highly recommended that old entity lists be periodically deleted. From the **XLDICTIONARY** facility, entity lists can be easily identified by their *ELS* entity type code.

Conclusion

You've just learned how to analyze your EXCELERATOR specifications for completeness, consistency, and accuracy. The **ANALYSIS** facility is among the most valuable of all EXCELERATOR facilities. Automated specification tools may be nice. But if they only allow analysts to automate their past mistakes, they will quickly outlive their usefulness. This is the facility in which we will likely see the most explosive growth of new features in all CASE products.

In its current version (1.7) EXCELERATOR provides extensive graphics analysis tools, mostly for data flow diagrams. These tools find the most common and serious errors associated with DFDs. Specifically, you learned how to print and use the Verification and Level Balancing reports. You also learned that EXCELERATOR has **Report Writer** and **Entity List** facilities that allow you to generate a variety of customized analytical reports against dictionary specifications. These reports can be saved for later use and shared with other projects.

You just about completed your tour and tutorial of EXCELERATOR facilities. The next lesson teaches you how to combine specifications, graphs, and analysis reports into single documents for printing.

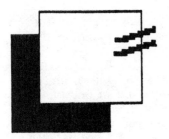

Lesson Nine: Packaging Documentation Using EXCELERATOR

The Demonstration Scenario

The record club project is coming along fine. Numerous diagrams and specifications have been created with EXCELERATOR. You've taken advantage of EXCELERATOR's ability to both maintain and analyze your specifications. Today, you received an electronic mail message from your boss:

> *" <your name>: I have scheduled a project quality assurance review meeting for your project. Please submit the following documentation to the committee: <message would insert a list of various graphs, records, elements, etc.>. You should organize the documentation as follows: <message would insert an outline>. In order to duplicate this documentation for all attendees, please have the master copy in my office by tomorrow morning, 8 A.M. Thanks!"*

This is the type of request, not all that uncommon, that turns your hair grey. Why? Producing special subsets of your specifications takes considerable time. In a typical project, even a small one, you generate large amounts of documentation. You usually develop a well conceived filing system (for instance, a tabbed three-ring binder) to store hard copies of specifications. Rarely is that filing system organized identical to the above request. Different requests for facts will need to include different subsets of the specifications in different sequences. Furthermore, you may want to include non-specifications such as management-oriented reports, cost/benefit data, project schedule data, and the like. Fortunately, you can prepare most of this supplemental information with computer aids such as word processors, spreadsheets, project managers, etc.

> *Wouldn't it be nice if you could tell EXCELERATOR to produce the desired subset of your specifications? And wouldn't it be nice if you could automatically tell EXCELERATOR how to organize the final document and then print it out with a single print command. Finally, wouldn't it be nice if you could also merge files from the word processor, spreadsheet, and the like into this document? You can do all of these things with EXCELERATOR!*

What Will You Learn in this Lesson?

Documentation has always been the *Catch 22* of the data processing industry. Most people will argue that thorough specification documents should be created, maintained, and made available throughout the life of the system. However, we may not be able to keep the specification data for non-current system projects on-line to EXCELERATOR (due to storage limitations). While we always want to maintain backups of an EXCELERATOR project for future use, we also want hard copy printouts of EXCELERATOR documentation. At least one set of all project data should be produced at the end of the project. The organization of this documentation may be subject to internal standards. Printing this sizable documentation could be unacceptably time-consuming.

In addition to printing a hard copy of the full set of documentation, there will usually be occasions on which you'll want to print a subset of all your project data. The format of the documentation will vary depending on its subsequent, intended use.

Finally, analysts frequently use other computer tools to enhance productivity, quality, and efficiency. These tools include the following:

o   Word processors. Word processors may be used to write memos, feasibility reports, summarize findings and conclusions, document specifications not possible with EXCELERATOR, present alternatives, propose solutions, and the like.

o   Spreadsheets. Spreadsheets may be used to document estimated costs, estimated benefits, and cost/benefit analyses.

o   Project managers. Project managers are used to set up, monitor, and control project schedules and resources. Most are based on PERT, CPM, Gantt, or some combination of those techniques.

o   Fourth-generation languages and applications generators. These tools allow analysts to prototype systems at an even higher level of sophistication than that offered by EXCELERATOR. Portions of such prototypes may (and probably should) find there way into our EXCELERATOR specifications.

EXCELERATOR provides a very useful **DOCUMENTATION** facility for combining various items created in the other facilities into a single document for printing. The document can include any of the following:

o   graphs of your choice
o   data dictionary outputs
o   screen and report design outputs
o   analyses reports
o   files from other software packages

An EXCELERATOR document is initially outlined as consisting of a series of named groups and fragments. The outline serves as an index into the actual files and records to be printed. The index is used to retrieve the files and print them in whatever order the outline specifies. Document outlines (graphs) can be saved for later use and shared between projects. (Sharing requires knowledge of the **XLD INTERFACE** facility.)

All of the above capabilities are provided through a Main Menu facility called **DOCUMENTATION**. Ideally, your organization should adopt standards or conventions for reusable **DOCUMENTATION** outlines.

After completing this lesson, you will be able to:

1.   Directly access your word processor and/or project manager via EXCELERATOR (assumption: availability of the software packages and proper installation of the packages into the EXCELERATOR environment -- ask your instructor or system manager).
2.   Create a graphical outline for a document.
3.   Check your outline for consistency with EXCELERATOR's documentation standards and print your document.

Exercise 9.1   Use EXCELERATOR's Interfaces to Your Computer's Word Processor and Project Manager (and Why You Might Want to Use Them)

The very nature of systems work requires numerous and various forms of written communication, including reports, memos, letters, and the like. Systems work also involves scheduling, control, and other project management functions. Such work can be enhanced greatly by the availability of word processing and project management software. EXCELERATOR does not include a word processing or project management software package. Why? There are numerous word processing and project management software packages on the market and many companies already make much use of one or more of these packages. Rather than dictate which word processor or project manager that must be used, EXCELERATOR provides the luxury of directly interfacing to the word processor or project manager software products of your choice.

In this exercise you'll learn how to access word processing and project management software packages from within EXCELERATOR's **DOCUMENTATION** facility (providing a word processor and/or project management package has been properly installed on your workstation by the instructor or systems manager - see *Guidelines for Setting Up the EXCELERATOR Environment*, immediately following the *Preface*). Since the access technique is the same for both, and since the word processor is the most common of the two packages, we'll demonstrate this facility with a word processor (again assuming that a word processor has been properly linked to EXCELERATOR).

Step 1. Log on to EXCELERATOR and get to the Main Menu. SELECT **DOCUMENTATION**. A Documentation Submenu should appear (see Figure 9-1). SELECT **Word Processing** from the submenu.

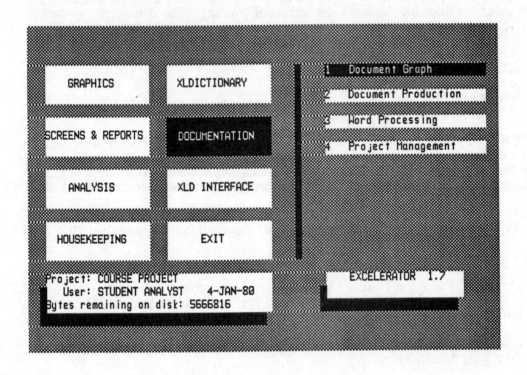

Figure 9-1

DOCUMENTATION Submenu Screen

Step 2. The EXCELERATOR screens should now be replaced by your word processor's screens. We want you to create the memo depicted in Figure 9-2. If you want to use your word processing document files in EXCELERATOR documents, you must consider the following caveats:

    o   According to InTech's manuals, you must open or save your word processor files as ASCII files. This means that the file cannot contain any of the special formatting codes (for instance, boldface, italics, underline, scripts, hanging indents, and the like). Consult your word processor's documentation to learn how to open or save your work as an ASCII file. On the other hand, we have discovered a way to retain special formatting for documents that will be merged into EXCELERATOR documents ...

    o   To retain special formatting, instead of saving the document file, print the document *to a file* (as opposed to printing the document to a printer). These print files can be incorporated into an EXCELERATOR document (the technique will be discussed in Exercise 9.2), complete with all formatting done by your word processor. This capability is not currently documented in the InTech manuals; however, we have tested it using *Microsoft WORD*.

Step 3. When you finished creating your document file, either print it as file (preferred, as noted above) or save it as an ASCII file (if the document is acceptable to you without any formatting, as noted above).

Step 4. Exit your word processor. This should automatically return you to EXCELERATOR's **DOCUMENTATION** submenu screen.

Gaining access to a project management software package is similar to word processing. You would SELECT **Project Management** from the Documentation Submenu.[1] Again, you may have to learn how to convert your project manager files to ASCII files or print them to a file if you intend to merge them into EXCELERATOR documents. Consult your project management software package's documentation for further details for techniques of conversion and/or printing to a file.

Exercise 9.2   Create a Graphical Outline of the Documentation Fragments That You Want to Merge

Organizing and packaging the specifications, and obtaining printouts can become quite tedious and time consuming if specifications are printed out one at a time or even in groups at a time. Fortunately, EXCELERATOR's **DOCUMENTATION** facility provides the capability of printing the entire document and its specifications, in the desired sequence or organization, in only three simple stages:

    1.   Create a graphical outline of the document you want to print.
    2.   Check the outline to ensure that EXCELERATOR can properly print the report.
    3.   Print the document.

This exercise covers the first stage. You begin by determining what specifications you would like to include in the document and how the document is to be organized. This same information must then be communicated or described to EXCELERATOR's **DOCUMENTATION** facility.

In this exercise you will learn how to create a graphical outline detailing an EXCELERATOR document's organization and content. The document you will produce is to contain several of the *EXCELERATOR* specifications you created in earlier lessons and a non-*EXCELERATOR* file (Figure 9-2) that you created in the previous exercise. If you did not create the ASCII file in the previous exercise, ask your instructor to provide you with one. Your instructor can explain the steps, in DOS, to copy a file into your EXCELERATOR project directory.

Let's suppose your instructor requires you to package and submit a formal document presenting a representative sample of the specifications you've completed from Lessons 3 through 9 of this tutorial. If

---

[1]   Although the submenu options read Word Processing and Project Management, you can actually link to any package. For instance, you may choose to substitute a spreadsheet for the project management package. Consult InTech's Installation Guide for details.

MEMORANDUM

TO: <instructor's name or course>

DATE: <date submitted>

FROM: <your name>

SUBJECT: Completed   samples   of   Record   Club   System
           Specifications

The following represents a representative number of
sample specifications completed for the Record Club
System.   The specifications were organized, packaged,
and printed using EXCELERATOR's DOCUMENTATION facility.

Figure 9-2

Memo to be Reproduced

you haven't completed each of those lessons, don't worry - you'll learn how to suppress printouts of those portions of the document that are not yet completed. For the sake of simplicity, let's assume the document is to be organized according to the tutorial's parts and lessons. A textual outline of the document follows[2]:

I.   Part Three Specifications
       A.   Word Processing document file (from Lesson 9)
       B.   DFD Description Screens (from Lesson 7)
II.   Part Two Specifications
       A.   A DFD (from Lesson 3)
       B.   DFDs (from Lesson 4)
       C.   Data modeling specifications (from Lesson 5)
             1.   The entity-relationship diagram
             2.   Record descriptions for all entities
       D.   Prototyping specifications (from Lesson 6)
             1.   A report design
             2.   A screen design

Let's communicate our document requirements to EXCELERATOR. The document's outline and content are described pictorially using **DOCUMENTATION's Document Graph** facility. Figure 9-3 represent the pictorial outline of the document we wish to produce. Let's recreate the graph.

Step 1. SELECT **DOCUMENTATION** from the Main Menu. Now SELECT **Document Graph** from the **DOCUMENTATION** submenu. An Action Keypad should appear. SELECT **Add**. You are now prompted to enter a name for the document graph. Type *RECORD CLUB SYSTEM SPECIFICATION* and press the Enter key.

Step 2. A drawing screen should now appear. This screen is quite similar to the drawing screens used in earlier lessons of the tutorial.

Step 3. SELECT OBJECT from the commands menu. Notice that a document graph contains two object types, **DOCGROUP** and **FRAGMENT**. These terms stand for *document group* and *document fragment*, respectively. They are defined as follows:

o   A *document group* refers to a logical unit within the document. For example, a document group may refer to a part or unit, chapter, section, or subsection of a document that you want to print.

o   A *document fragment* refers either EXCELERATOR graphs or specifications, or non-EXCELERATOR files to be included in the document. For example, a document fragment may refer to non-EXCELERATOR ASCII files, data flow diagrams, dictionary descriptions, report designs, and other EXCELERATOR produced outputs.

On Figure 9-3, document groups are depicted as a stack of overlapping rectangles. Document fragments are depicted as a single rectangle.

Before continuing with the exercise, you should understand how a document graph is constructed and interpreted (referring to Figure 9-3 as an example). Just as a textual outline communicates a document's hierarchical organization and sequencing, so does a document graph. The left-most object on the document graph is referred to as the *root object* and represents the document as a whole. EXCELERATOR processes the graph by reading from left-to-right, top-to-bottom. The print path through our graph is shown in Figure 9-4. The top-left represents the start of the document and the bottom-right represents the end of the document.

Take a moment to study Figure 9-3 or 9-4. Then compare it to the textual outline. They are equivalent! Every textual outline has a corresponding document group (GRP). Eventually, lowest level groups have a corresponding fragment (FRG) entity that points to the actual files and specifications. The equivalence is summarized as follows:

---

2   The sequencing of the parts and lessons within the outline are not important. The authors chose this organization for demonstration purposes only. You or your instructor may modify the organization after completing reading this exercise.

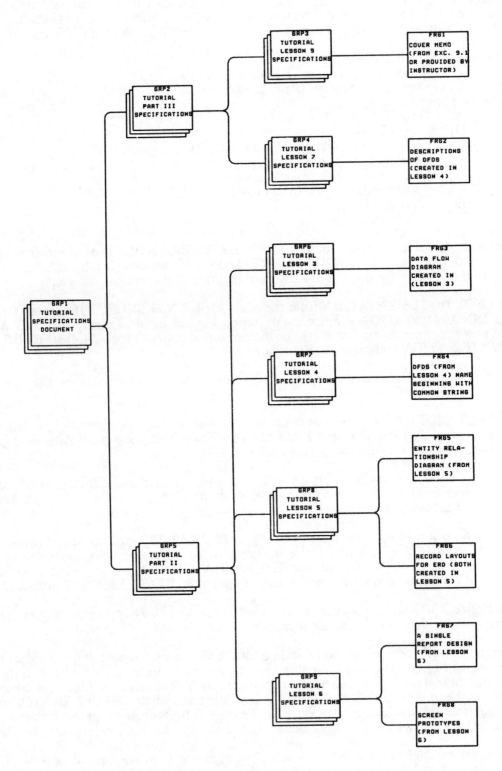

Figure 9-3

A Document Graph: A Pictorial Outline of a Document

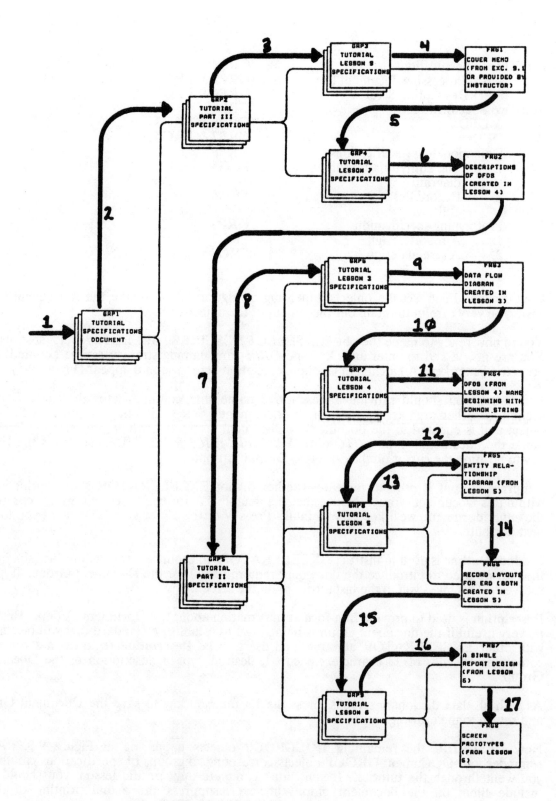

Figure 9-4

Print Path Through the Document Graph

Textual Outline Item	Graphical Outline Object
Document	GRP1
I. Part Three Specifications	GRP2
A. Word Processing document file	GRP3 & FRG1
B. DFD Description Screens	GRP4 & FRG2
II. Part Two Specifications	GRP5
A. A DFD	GRP6 & FRG3
B. DFDs	GRP7 & FRG4
C. Data modeling specifications	GRP8
1. The entity-relationship diagram	FRG5
2. Record descriptions for all entities	FRG6
D. Prototyping specifications	GRP9
1. A report design	FRG7
2. A screen design	FRG8

Step 4.   Now draw and connect the objects appearing in Figure 9-3. To make the subsequent steps easier, be very careful to retain our placement of the objects.

Step 5.   You're now ready to name the objects. SELECT **DESCRIBE**. SELECT the left-most object. You are now asked to enter an ID. Type *GRP1* (any naming strategy can be adopted) and press the Enter key. A Document Group Description Screen should appear (Figure 9-5).

For **Label**, you should provide a descriptive name that communicates the unit, chapter, section, or other organization unit of the document. Since this object represents the root object and is equated to the document or document graph as a whole, let's use the label to name the document. Type *TUTORIAL SPECIFICATIONS DOCUMENT*. Press the Enter key to position the cursor on the **Suppress Output** attribute.

Suppress Output is used to indicate whether or not EXCELERATOR should print items within this Document Group. Notice that the default is *N* for *no*. Since this group represents the *entire* document, we'll keep the default. Press *N* or the Enter key. This takes you to the next attribute.

Produce Outline is used to instruct EXCELERATOR to produce a textual outline (similar to the one we used to introduce the document requirements) for the Document Group. Type *Y* for "yes". This takes you to the last attribute on the screen.

Description is used to provide free-format information about the Document Group. This can be very useful if the document graph is being used to establish a standard that will be shared with other EXCELERATOR accounts and users. The **Description** region is a scroll area containing 60 lines of 72 characters each. If desired, type a description of the Document Group.

As with all data dictionary screens, press the F3 function key to save the Document Group and return to the drawing screen.

Step 6.   Now **DESCRIBE** the remaining **DOCGROUP** objects appearing in Figure 9-3. As a reminder, our DOCument GROUP objects correspond to groups of specifications created as you went through the tutorial. If you didn't complete any specific lesson, you should still include them on the document graph; however, suppress the actual printing of those specifications or files (as described in Step 5, **SUPPRESS OUTPUT**).

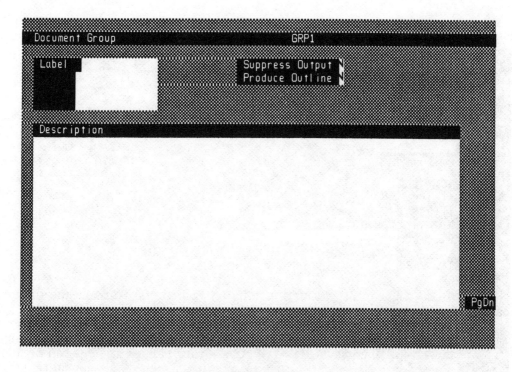

Figure 9-5

A Document Group Description Screen

Step 7.   Now let's label the Document Fragment (**FRAGMENT**) objects appearing in Figure 9-3. SELECT **DESCRIBE**. SELECT the top Document Fragment object on the screen. Type the ID *FRG1* (any naming strategy can be adopted) and press the Enter key. A Document Fragment Description Screen should appear (Figure 9-6). The description contains six attribute fields.

The **Label** attribute is used to provide a brief description of the Fragment.   Recall that a Fragment refers to one or more files or XLDictionary entities. Type a brief description and press the Enter key.

Fragment Type is used to identify the type of information that is represented by the Fragment. A three-letter code is used to indicate the type. The codes correspond, in part, to the entity types you learned in Lesson 7. In the interest of completeness, Figure 9-7 describes all valid fragment types. Since our first document comes from a non-EXCELERATOR word processing package, type *TXF*.

Output Action is used to enter one of three possible output options:

o   "*L*" means *list* and is used to output only the names of desired items.
o   "*D*" means *describe* and results in the output of all attributes of every data dictionary record selected. Obviously, this results in the most detailed output option.
o   "*I*" for *image/invoke*, is used to obtain output for all selected graphs and all selected non-EXCELERATOR files.

For your first fragment, the correct response is *I* (since it came off a word processor, not EXCELERATOR).

181

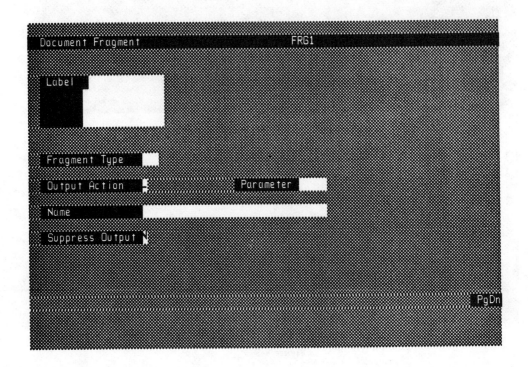

Figure 9-6

A Document Fragment Description Screen

FRAGMENT TYPES AND THEIR MEANING AND USAGE

All entity types including:

```
DAE for Data Entity
DAF for Data Flow
DAR for Data Relationship
DAS for Data Store
DCG for Document Graph
DMD for Data Model Diagram
DNA for Data N-Ary Relationship
DOC for Document Group
DOF for Document Fragment
ELE for Element
ELS for Entity List
ERA for Entity-Relationship Diagram
EXT for External Entity
FUN for Function
MOD for Module
PGC for Presentation Graph Connection
PGO for Presentation Graph Object
PRC for Process
PRG for Presentation Graph
REC for Record
RED for Report Design
REP for Report
SCD for Screen Design
SDE for Screen Data Entry
SDR for Screen Data Report
SDV for System Device
SGC for Structure Graph Connection
STC for Structure Chart
STD for Structure Diagram
TAB for Table of Codes
USR for User
```

Figure 9-7(a)

Fragment Codes (Part 1)

XLD relationship types:

Only two XLD relationship types are
permitted. The two possible codes are
are RCE and RCR. RCE stands for the
relationship RECord Contains ELEment,
wherein all elements of a record are
referenced. RCR stands for the
relationship RECord Contains RECord,
wherein all sub-records for a record are
referenced.

Graph ANALYSIS reports:

All graph ANALYSIS reports. Permissible
codes and their meanings include:

* SAR for Analysis Report
* GER for Graph Explosion
Report
* LLR for Level Balancing
Report
* SVR for Verification
Report

Non-EXCELERATOR files:

ASCII or print files from non-EXCELERATOR
software packages must use the code TXF.
It is the EXCELERATOR user's
responsibilities to get these files into
their EXCELERATOR project directory
(using DOS). Consult your instructor or
system manager if you need to learn how
to COPY files into your project
directory.

Expansion document graphs:

Any Expansion Document Graph. An
Expansion Document Graph is a Document
Graph that can be appended to another
Document Graph. Expansion Document Graph
must be given the Fragment Type code EXP.

Figure 9-7(b)

Fragment Codes (Part 2)

Parameter is only used for fragments whose **Fragment Type** is either *SCD* (Screen Design), *LLR* (Level Balancing Report), or *GER* (Graph Explosion Report) *and* whose **Output Action** is either *D* (Describe) or *I* (Image). Essentially, this Parameter attribute allows you to provide additional output instructions for required fragment types:

o   For Screen Designs, two one-letter codes must be provided. The first code is a Screen Header information option and must be either an *S* for Summary, *D* for Detailed, or *N* for None. The second code is a Field Description option which must also be either an *S* (Summary), *D* (Detailed), or an *N* (None).

o   For Level Balancing Reports, only one parameter must be specified, the number of levels to balance.

o   For Graph Explosion Reports, you must type the graph type (such as DFD for data flow diagrams) and a number corresponding to the number of explosion levels to be included on the report. Explosions levels can be 0-9 levels.

Since the current Document Fragment you are DESCRIBing is of Fragment Type *TXF*, you can skip this attribute. Press the Tab or Enter key.

The **Name** attribute is used to provide a name or name range for those items (non-EXCELERATOR ASCII output files or EXCELERATOR output files) that are included in the Document Fragment. Entries can either be a specific item name, a wildcard for obtaining all occurrences of the specified Fragment Type (a single asterisk), the name of an entity list that points to specific entities, or a prefix followed by a wildcard to obtain all occurrences of the specified Fragment Type whose name begins with the same character string (for example, EMPLOYEE*). Unfortunately, there is no selector list available for completing this attribute. Therefore, you should know the item names before electing to DESCRIBE a Document Fragment. If necessary, you may use the XLDICTIONARY facility to determine item names. Recall the name of the file you created in Exercise 9.1 (or the name of the file provided by your instructor). Type the full name of the file (note: the authors' suggested name was MEMO.PRN), and press the Enter key.

Suppress Output is used the same as with DESCRIBing Document Groups. This attribute is used to prevent the actual printing of the Document Fragment. You may find this attribute useful in suppressing the output of those Document Fragments that you have not yet completed. If you did Exercise 9.1, type *N* for *no*. Otherwise, type *Y* for *yes*.

Next, press the PgDn key. An additional Document Fragment Description Screen appears. The cursor is positioned at a **Description** attribute field. This attribute is used to provide a brief description of the Document Fragment. If desired, type in a brief description.

Press F3 to return to the Document Graph Drawing Screen.

Step 8.   Now **DESCRIBE** the remaining Document Fragments appearing on the document graph. For ease and understanding, Figure 9-8 provides the Descriptions for all Document Fragments appearing in Figure 9-3. It is important to note that you may need to make appropriate adjustments to the Suppress Output entries in the event that you haven't actually completed the items referenced by the Document Fragment. You may also need to change the Name attribute entries to reflect the actual item names you may have assigned.

Step 9.   SELECT **EXIT** and then **SAVE** from the command menu. This will return you to the Document Graph Action Keypad. Now SELECT **EXIT** to return to the Main Menu and **DOCUMENTATION** Facility Menu.

In the next exercise you'll learn how to verify the accuracy of your Document Graph and print your document. But first, you should recognize the value of the document graph itself. The Document Graph can easily be customized to your preference. For example, the document graph can be modified to reflect a different organization or to include additional Document Groups and Document Fragments for including other desired specifications. Alternatively, the document can be modified to delete or suppress the printing

```
DATE: 25-MAR-87      DOCUMENT FRAGMENT - OUTPUT                    PAGE    1
TIME: 16:35          NAME: *                                  EXCELERATOR  1.7

TYPE Document Fragment                   NAME FRG1

   Label COVER MEMO
         (FROM EXC. 9.1
         OR PROVIDED BY
         INSTRUCTOR)

   Fragment Type   TXF

   Output Action   I                 Parameter

   Name            MEMO.DOC

   Suppress Output N

   Description
   THIS DOCUMENT FRAGMENT REFERENCES A NON-EXCELERATOR ASCII FILE.  THE
   FILE MAY HAVE BEEN CREATED FROM FIGURE 9-2 IN EXERCISE 9.1.
   ALTERNATIVELY THE FILE MAY HAVE BEEN PROVIDED BY THE INSTRUCTOR.  IN
   EITHER CASE, THE ASCII FILE MUST RESIDE IN THE STUDENT'S EXCELERATOR
   PROJECT DIRECTORY.

   Modified By   STUDENT ANALYST   Date Modified  800104    # Changes  2
   Added By      STUDENT ANALYST   Date Added     800104
   Last Project  COURSE PROJECT
   Locked By                       Date Locked    0         Lock Status
```

```
DATE: 25-MAR-87      DOCUMENT FRAGMENT - OUTPUT                    PAGE    2
TIME: 16:35          NAME: *                                  EXCELERATOR  1.7

TYPE Document Fragment                   NAME FRG2

   Label DESCRIPTIONS
         OF DFDS
         (CREATED IN
         LESSON 4)

   Fragment Type   DFD

   Output Action   D                 Parameter

   Name            *

   Suppress Output N

   Description
   THIS DOCUMENT FRAGMENT REFERENCES XLDICTIONARY DESCRIPTIONS FOR DATA
   FLOW DIAGRAMS.  THE DESCRIPTIONS WERE CREATED IN LESSON 7 OF THE
   TUTORIAL.

   Modified By   STUDENT ANALYST   Date Modified  800104    # Changes  1
   Added By      STUDENT ANALYST   Date Added     800104
   Last Project  COURSE PROJECT
   Locked By                       Date Locked    0         Lock Status
```

```
DATE: 25-MAR-87      DOCUMENT FRAGMENT - OUTPUT                    PAGE    3
TIME: 16:35          NAME: *                                  EXCELERATOR  1.7

TYPE Document Fragment                   NAME FRG3

   Label DATA FLOW
         DIAGRAM
         CREATED IN
         (LESSON 3)

   Fragment Type   DFD

   Output Action   I                 Parameter

   Name            SUBSCRIPTION PROCESSING 1ST DRFT

   Suppress Output N

   Description
   THIS DOCUMENT FRAGMENT REFERENCES THE DATA FLOW DIAGRAM CREATED IN
   LESSON 3 TO LEARN THE BASICS OF EXCELERATOR'S GRAPHICS FACILITY.  ONLY
   THE DIAGRAM ITSELF IS TO BE PRINTED.

   Modified By   STUDENT ANALYST   Date Modified  800104    # Changes  1
   Added By      STUDENT ANALYST   Date Added     800104
   Last Project  COURSE PROJECT
   Locked By                       Date Locked    0         Lock Status
```

Figure 9-8(a)

Document Fragment Descriptions (Part 1)

DATE: 25-MAR-87 DOCUMENT FRAGMENT - OUTPUT PAGE 4
TIME: 16:35 NAME: * EXCELERATOR 1.7

TYPE Document Fragment NAME FRG4

Label DFDS (FROM
LESSON 4) NAME
BEGINNING WITH
COMMON STRING

Fragment Type DFD

Output Action I Parameter

Name RECORD CLUB SYSTEM*

Suppress Output N

Description
THIS DOCUMENT FRAGMENT REFERENCES THOSE DFDS CREATED IN LESSON 4 WHOSE NAME BEGINS WITH THE CHARACTER STRING "RECORD CLUB SYSTEM". IT SHOULD BE POINTED OUT THAT, IN OUR EXAMPLE, THERE WILL BE TWO ACTUAL DFDS AND A HIERARCHY CHART. THAT THE HIERARCHY CHART WAS CREATED USING THE THE DATA FLOW DIAGRAM FACILITY OF GRAPHICS.

Modified By	STUDENT ANALYST	Date Modified	800104	# Changes	1
Added By	STUDENT ANALYST	Date Added	800104		
Last Project	COURSE PROJECT				
Locked By		Date Locked	0	Lock Status	

DATE: 25-MAR-87 DOCUMENT FRAGMENT - OUTPUT PAGE 5
TIME: 16:35 NAME: * EXCELERATOR 1.7

TYPE Document Fragment NAME FRG5

Label ENTITY RELA-
TIONSHIP
DIAGRAM (FROM
LESSON 5)

Fragment Type ERA

Output Action I Parameter

Name RECORD CLUB SYSTEM DATA MODEL

Suppress Output N

Description
THIS DOCUMENT FRAGMENT REFERENCES THE ENTITY RELATIONSHIP DIAGRAM FOR THE RECORD CLUB SYSTEM. THE ERD WAS CREATED IN LESSON 5 OF THE TUTORIAL.

Modified By	STUDENT ANALYST	Date Modified	800104	# Changes	2
Added By	STUDENT ANALYST	Date Added	800104		
Last Project	COURSE PROJECT				
Locked By		Date Locked	0	Lock Status	

DATE: 25-MAR-87 DOCUMENT FRAGMENT - OUTPUT PAGE 6
TIME: 16:35 NAME: * EXCELERATOR 1.7

TYPE Document Fragment NAME FRG6

Label RECORD LAYOUTS
FOR ERD (BOTH
CREATED IN
LESSON 5)

Fragment Type REC

Output Action D Parameter

Name *

Suppress Output N

Description
THIS DOCUMENT FRAGMENT REFERENCES RECORD LAYOUTS DEFINED IN THE DICTIONARY. MOST OF THE RECORD LAYOUTS WERE CREATED AS EXPLOSIONS OF ENTITIES APPEARING ON THE ENTITY RELATIONSHIP DIAGRAM DRAWN IN LESSON 5 OF THE TUTORIAL.

Modified By	STUDENT ANALYST	Date Modified	800104	# Changes	1
Added By	STUDENT ANALYST	Date Added	800104		
Last Project	COURSE PROJECT				
Locked By		Date Locked	0	Lock Status	

Figure 9-8(b)

Document Fragment Descriptions (Part 2)

TYPE Document Fragment   NAME FRG7

Label A SINGLE
      REPORT DESIGN
      (FROM LESSON
      6)

Fragment Type   RED

Output Action   I   Parameter

Name   ORDER RESPONSE REPORT DESIGN

Suppress Output N

Description
THIS DOCUMENT FRAGMENT REFERENCES A SPECIFIC REPORT DESIGN FROM LESSON
6 OF THE TUTORIAL. SPECIFICALLY, IT REFERENCES THE REPORT DESIGN
DEPICTED IN FIGURE 6-1 OF LESSON 6.

Modified By	STUDENT ANALYST	Date Modified	800104	# Changes	1
Added By	STUDENT ANALYST	Date Added	800104		
Last Project	COURSE PROJECT				
Locked By		Date Locked	0	Lock Status	

TYPE Document Fragment   NAME FRG8

Label SCREEN
      PROTOTYPES
      (FROM LESSON
      6)

Fragment Type   SCD

Output Action   I   Parameter SS

Name   ORDER MENU

Suppress Output N

Description
THIS DOCUMENT FRAGMENT REFERENCES SCREEN PROTOTYPES CREATED IN LESSON 6
OF THE TUTORIAL. SCREEN HEADER AND FIELD DESCRIPTION INFORMATION TO
ACCOMPANY THE SCREEN IMAGE ARE TO CONTAIN SUMMARY INFORMATION.

Modified By	STUDENT ANALYST	Date Modified	800104	# Changes	1
Added By	STUDENT ANALYST	Date Added	800104		
Last Project	COURSE PROJECT				
Locked By		Date Locked	0	Lock Status	

TYPE Document Fragment   NAME TEST FRAG 1

Label VERIFICATION
      REPORT

Fragment Type   VER

Output Action   L   Parameter

Name

Suppress Output N

Description

Modified By	STUDENT ANALYST	Date Modified	870128	# Changes	1
Added By	STUDENT ANALYST	Date Added	870128		
Last Project	COURSE PROJECT				
Locked By		Date Locked	0	Lock Status	

Figure 9-8(c)

Document Fragment Descriptions (Part 3)

of certain Document Groups and Document Fragments. However, the ultimate value of the Document Graph is realized through sharing. For example, your instructor could have created the Document Graph and made it available to all students completing the tutorial. This would ensure consistency. Many companies have adopted reporting standards which can be communicated via sharing a common Document Graph. Making a Document Graph available to other EXCELERATOR users is accomplished through the XLD INTERFACE facility (discussed in Lesson 10).

Exercise 9.3   Verify a Graphical Outline and Print the Documentation

If you've completed Exercise 9.2 you now have a Document Graph that communicates the organization and content of a fairly large document. Your graph contained numerous Document Groups and Document Fragments. You should realize that it is quite reasonable to assume that mistakes could have been made. In Lesson 8 of the tutorial, you learned about EXCELERATOR's **ANALYSIS** facility and how it can be used to ensure accuracy, consistency, and completeness of specifications. EXCELERATOR also provides a facility called **Document Production** for checking Document Graphs to ensure their accuracy and completeness.

In this exercise you'll learn how to use the **Document Production** facility from the **DOCUMENTATION** Facility Menu to verify a Document Graph and produce a printout of your document. Let's first verify and print the Document Graph you created in the last exercise.

Step 1.   SELECT **Document Production** from the **DOCUMENTATION** Facility Menu. You are now provided with three Document Production options (Figure 9-9). Now SELECT **Doc Graph**. An Action Keypad should appear (Figure 9-10). The **Doc Graph** facility allows you to obtain a list of existing Document Graphs, verify a Document Graph to check its accuracy, and obtain a file or printer output of the document. SELECT **Verify** from the Action Keypad. You are now prompted for a Document Graph name. Enter the name of the Document Graph or press the Enter key and SELECT the Document Graph from the selector list that appears.

Step 2.   EXCELERATOR is now verifying your Document Graph. EXCELERATOR first checks the Document Graph to ensure it is formatted appropriately. Document Fragments appearing on the Document Graph are then checked for accuracy. Document Fragments are checked to ensure that valid codes for Fragment Types and Parameters were provided when they were **DESCRIBE**d. In addition, EXCELERATOR checks to make sure that all items referenced in the Name attribute field of the Document Fragment Description do actually exist in your project directory. Any errors that were detected should be displayed across the screen. Such errors must be corrected in order for the document to be successfully printed.

Option.   If errors appeared, those errors can be corrected *without* having to go into the **DOCUMENTATION Document Graph** facility. The **Doc Group** and **Doc Fragment** options on the Action Keypad can be used to make modifications to a Document Group or Document Fragment containing errors. If EXCELERATOR detected errors, you should press a non-arrow key and then SELECT **Doc Group** or **Doc Fragment**. An Action Keypad would appear offering numerous options for maintaining Document Groups or Fragments.

If your Document Graph contained errors related to a Document Fragment, you should SELECT **Modify**, enter the name, and modify the description that appears. A special Action Keypad option is available for document fragments (Figure 9-11). A separate **Verify** option exists for retesting the particular Document Fragment to see the errors were corrected. In other words, there's no need to re-**Verify** the *entire* Document Graph. Note that verifications can only be performed upon a Document Graph as a whole or upon specific Document Fragments (not upon Document Groups).

Step 3.   After and only after the Document Graph has been successfully verified, can you obtain a printout of your document (repeat Steps 1 and 2 if necessary). Once your document has been successfully verified, SELECT **Doc Graph**. Now SELECT **Execute** from the Action Keypad

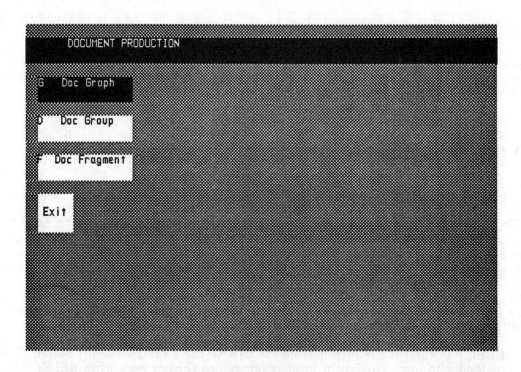

Figure 9-9

Document Production Submenu Screen

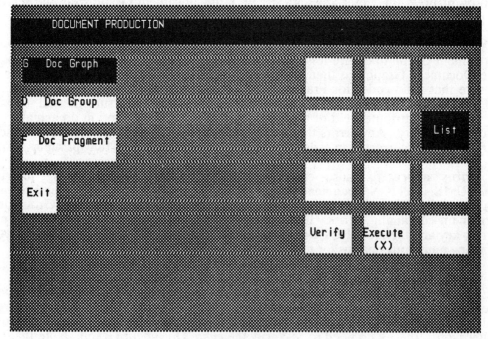

Figure 9-10

Document Graph Action Keypad Screen

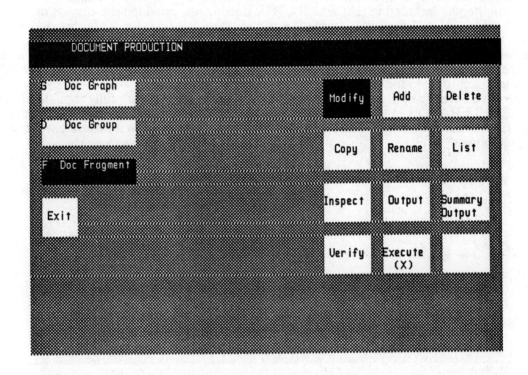

Figure 9-11

Verify Action Keypad Option for Document Fragment

that appears. You should now be prompted for a Document Graph name. Type the name of your Document Graph or press the Enter key and SELECT the Document Graph. EXCELERATOR now prompts you for where to send the output. SELECT Printer.

Don't be too anxious about receiving your output. A delay may occur while EXCELERATOR locates the appropriate files. Also, you shouldn't become concerned if the screen changes periodically. EXCELERATOR will notify you when printing is completed or is interrupted. Be patient.

Step 4.   After your document is printed, you may wish to **Exit** and backup your work.

Conclusion

You've just learned how to successfully package your EXCELERATOR specifications into a very professional document for presentation to your instructor. You've learned how to use the **DOCUMENTATION** facilities to interface with your word processor and project manager. The files from these products can then be included in EXCELERATOR documents, provided the caveats described in this lesson are followed. You have also learned a three step process for defining and printing an EXCELERATOR document that contains various types of EXCELERATOR files, dictionary specifications, and reports, as well as non-EXCELERATOR files. First, you must create a **Document Graph** that pictorially outlines your document. Second, you verify the correctness of the outline. Third, you print the document according to the verified outline.

You now have a wealth of knowledge about EXCELERATOR. You are fully capable of introducing and/or using EXCELERATOR in your future projects (and employers). However, if you want to become an even better EXCELERATOR user and learn more about the product, you'll certainly want to complete Lesson 10, which follows.

Lesson Ten: Where Do You Go From Here?

What Will You Learn in this Lesson?

This tutorial has brought you up to speed with EXCELERATOR. The purpose of this lesson is to provide you with some direction for your continued EXCELERATOR education. After completing this lesson you will be able to:

1. Describe features of EXCELERATOR which were not covered in this tutorial but which you may fund useful in some of your subsequent EXCELERATOR projects.
2. Describe available EXCELERATOR documentation and how it will assist your continued education.
3. Describe *EXCELERATOR/RTS*, an alternative version of EXCELERATOR, used for real-time applications.
4. Describe *CUSTOMIZER*, a value-added product for EXCELERATOR, used to customize the base product.
5. Describe the *VDAM Mainframe Interface*, another value-added product.
6. Describe the *XL Group* and its opportunities.

Additional EXCELERATOR Facilities

We've provided 80-20 coverage of EXCELERATOR's facilities in this tutorial - that is, it would take an 80% larger book to teach you the remaining 20% of EXCELERATOR. Fortunately, you've learned enough about EXCELERATOR to easily learn additional features and techniques from InTech's own Reference Manual and User's Guide (there is no substitute for any software package's detailed documentation). What we want to do in this section is to give you some idea of the capabilities that we omitted from this tutorial. This, in essence, will serve as a directory of additional assistance that you'll find in InTech's own manuals.

Other Graphics Features. In Part Two you learned how to use the **GRAPHICS** facility to draw data flow diagrams and entity-relationship diagrams. EXCELERATOR is also capable of drawing the following graph types:

o **Yourdon-style Data Flow Diagrams.** It bears repeating that the data flow diagrams printed in this book use the Gane and Sarson symbol set. If your user account has the security privilege, *Project Manager*, you can switch to the Yourdon (and DeMarco) symbol set. From the Main Menu, you would SELECT HOUSEKEEPING, **Project Manager**, and **Modify**. Then type or

SELECT your project name and change the **DATA FLOW DIAGRAMS** attribute to *Y* (for Yourdon) or *G* (for Gane and Sarson).

o   **Merise-style Entity-Relationship Diagrams.** Figure 10-1 illustrates the Merise symbol set for a entity-relationship data model. The technique for switching to this symbol set is the same as for changing to Yourdon data flow diagrams except that you set the **DATA MODELING** attribute to *M* (for Merise).

Note that both styles of entity-relationship diagrams can be exploded out of the data stores from data flow diagrams. This means that you can easily integrate data flow and data modeling approaches with EXCELERATOR.

o   **Structure Charts.** If you are familiar with *Structured Analysis*'s companion methodology, *Structured Design* (Yourdon approach), you can draw structure charts (Figure 10-2). Structure charts show the hierarchical, modular construction of proposed computer programs. The modules can be illustrated as *to be coded* (rectangles) and *previously defined/library* (rectangles with side bars). Decisions, looping, and parameter passing are also accommodated. Structure charts can be exploded out of processes on data flow diagrams.

o   **Data Model Diagrams.** This is an alternative data modeling approach for those who don't want to use the entity-relationship diagrams. You can also combine it with entity-relationship diagrams (e.g. use entity-relationship diagrams to model the structure of data stores and data bases, and use data model diagrams to model the structure of data flows, from which they can be exploded.

Figures 10-3 and 10-4 demonstrate the alternative ways to depict one-to-many connections (double arrows versus "crows' feet").

o   **Jackson-style Structure Diagrams.** If you are familiar with Jackson's version of *Structured Design* for computer programs, you will recognize structure diagrams (Figure 10-5). Note that this diagrammatic technique can be used to implement a hierarchical version of Warnier/Orr Diagrams (otherwise omitted from EXCELERATOR). Both Jackson and Warnier/Orr use data-oriented program design strategies.

o   **Presentation Graphs.** This is a superset of the standard systems and program flowcharting templates with which you may be familiar. It also supports the Joint Application Design (JAD) methodology. With **Presentation Graph**, you can draw graphs similar to Figure 10-6.

Other XLDictionary Features. You have learned virtually all of the functionality built into EXCELERATOR's **XLDICTIONARY** facility. We stressed that facility since it is the heart and soul of EXCELERATOR's potential as a productivity and quality enhancing tool. Keep in mind that all entities can be recorded in the dictionary for later reference and reuse.

Other Screens & Reports Features. If you covered Lesson 6, you learned all of the facilities available in **SCREENS & REPORTS**.

Other Documentation Features. You have at least learned how to use all of **DOCUMENTATION**'s facilities. We hope you also learned that you can use your word processor and project manager from within **DOCUMENTATION** (but that depends on how your workstation was set up).

Other Analysis Features. There were two graph analysis reports that we didn't cover. We should at least bring them to your attention. The first is called **Graph Explosion**, and is the only graph analysis report that works with anything other than data flow diagrams. The report serves as a check on completeness and consistency of explosions for a graph. For instance, this report will inform you of explosions that have been described to the dictionary but not yet drawn.

The second graph analysis report is called the **Analysis Report**. This report can be very useful once you have described all DFD data flows down to the data element level. Essentially, the report documents, for a specified process, all the input data flows (including file retrievals) and their contents, and the all of the output data flows (including file updates) and their contents. Consequently, this report helps the analyst

194

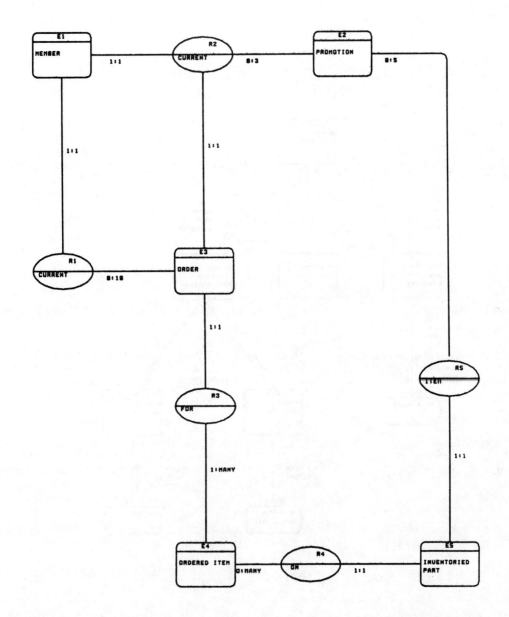

Figure 10-1

Merise Entity-Relationship Diagram

195

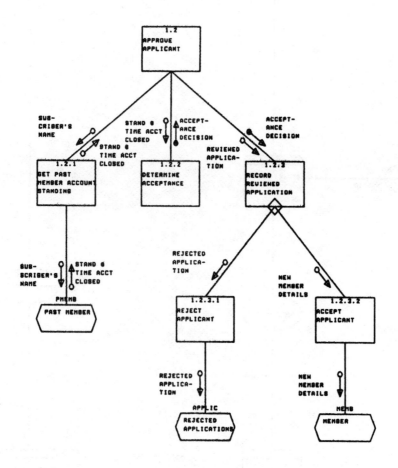

Figure 10-2

Yourdon Structure Chart

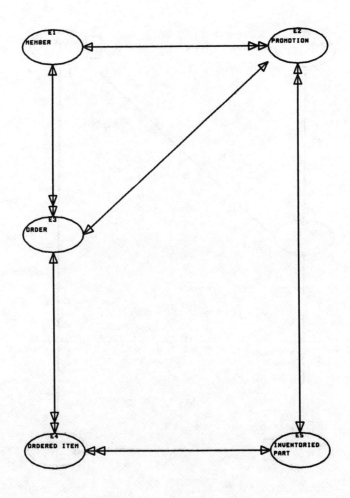

Figure 10-3

Data Model Diagram (Version 1)

197

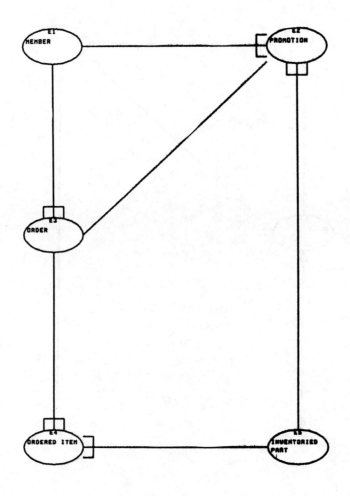

Figure 10-4

Data Model Diagram (Version 2)

198

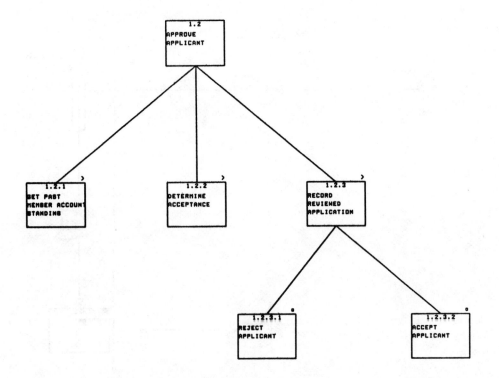

NOTE: THIS SAMPLE STRUCTURE DIAGRAM REPRESENTS THE EQUIVA-
LENT HIERARCHICAL STRUCTURE DEPICTED IN THE FORM OF A
STRUCTURE CHART IN FIGURE 10-2. COMPARE THE TWO FIGURES.
NOTICE THAT THE JACKSON-STYLE USES SYMBOLS (APPEARING ABOVE
SPECIFIC PROCESS FUNCTIONS) TO COMMUNICATE PROCESSING LOGIC.
REFER TO THE EXCELERATOR USER GUIDE FOR A FULL DESCRIPTION
OF SYMBOL TYPES AND THEIR MEANINGS.

Figure 10-5

Jackson Structure Diagram

PROPOSED ORDER PROCESSING FLOWCHART

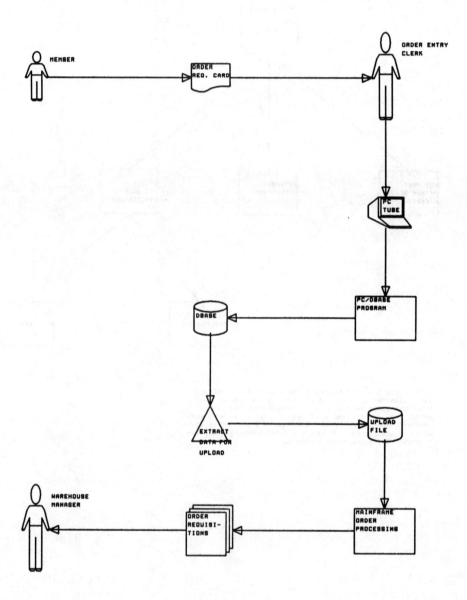

Figure 10-6

Presentation Graph

ensure that the inputs are sufficient to produce the outputs. Frequently, any such problems go unnoticed until programming. They are much more expensive to fix at that stage of development.

We cannot overemphasize that you become more familiar with entity lists. InTech's User Guide provides some interesting and well written explanations of the power of entity lists to create customized analyses of your system specifications. Most users have little appreciation for how easy it is to create and use customized reports.

The XLD Interface Facility. We haven't spent too much time on this facility. In the last lesson, you learned how to **Import** files from other software packages. In reality, **XLD INTERFACE**'s full potential lies in its ability to control data sharing across projects. Data sharing can occur on one workstation, or across multiple workstations. This permits the work for a single project to be distributed across several workstations or accounts with complete control over a common dictionary. **XLD INTERFACE**'s facilities are described as follows:

o **Export.** This is the inverse of **Import**. You might use export to transfer files to a diskette which can then be **Import**ed to another workstation's project.

   Figure 10-7 illustrates a typical export screen that creates a file call *MEMBER.EXP*. That file can then be **Export**ed to any other project using the facility described above.

o **Import.** This is used to import files from other EXCELERATOR project or sub-project directories. You can import entire project directories, all entities referenced on a named entity list, or all entities that fulfill selection criteria that you specify. Consequently, EXCELERATOR users can share common data elements, tables of codes, document fragment templates, customized analysis reports, etc.

o **Lock.** This facility locks selected entities to prevent other accounts from changing those entities. The audit attributes in all data dictionary screens indicate the lock status and who locked the entity. Consequently, you would know whom to go to if you wished to determine why the entity was locked and get approval to change the entity.

o **Unlock.** The inverse of lock.

o **Export & Lock.** A combination of the above.

o **Import & Lock.** A combination of the above.

Other Housekeeping Facilities. There are two facilities that are probably unfamiliar to you. The term **Project Manager** can be confusing. In the EXCELERATOR environment, it does not refer to support tools for project management, scheduling, resource management and control. Instead, it is the facility used to create, modify, and delete project accounts and defaults. **System Manager**, another facility, is used to create, modify, and delete user accounts and system defaults. Both **Project Manager** and **System Manager** require special access privileges (normally established by the workstation manager(s)).

Where Do You Learn About These Additional Facilities? InTech provides complete documentation for their products. For EXCELERATOR, you should be aware that the following manuals are shipped with every copy of the product:

o *EXCELERATOR Reference Guide.* This manual, organized by facility, is the final reference word for all your questions. Since you have now completed this tutorial, you should be able to easily locate and read instructions from this manual.

o *EXCELERATOR User Guide.* In our opinion, this is your next reading assignment. This excellent manual will provide additional tips on how to fully exploit the power of EXCELERATOR. We strongly recommend you carefully study the excellent coverage of entity lists and how to more fully exploit them.

o *EXCELERATOR Tutorial.* EXCELERATOR provides its own tutorial. It is structured somewhat differently than ours; however, it is well written and may prove useful as an additional source of training or exercises.

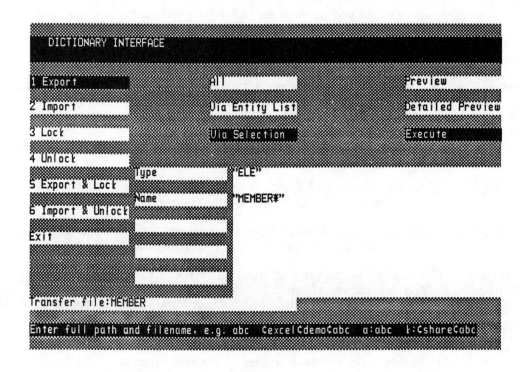

DICTIONARY INTERFACE

1 Export   All   Preview
2 Import   Via Entity List   Detailed Preview
3 Lock   Via Selection   Execute
4 Unlock
5 Export & Lock   Type   "ELE"
6 Import & Unlock   Name   "MEMBER*"
Exit

Transfer file:MEMBER
Enter full path and filename, e.g. abc ¢excel¢demo¢abc a:abc k:¢share¢abc

Figure 10-7

Export Screen

EXCELERATOR/RTS: A Specialized Version of the Standard Package

This tutorial was designed for EXCELERATOR's premier, basic package. Alternatively, there is a package called *EXCELERATOR/RTS*. RTS stands for *Real Time Systems*. The package contains a subset of EXCELERATOR's basic tools plus some additional tools specifically geared to the needs of analysts and engineers who develop real-time systems (e.g. computer-assisted manufacturing, commercial aviation, manufacturing process control, telecommunications, operating systems, etc.). It is largely based on the real-time analysis and design methodologies suggested by Derek Hatley, Stephen Mellor, and Paul Ward.

Traditional data flow and data structure analysis provided in the the version of EXCELERATOR that you have just learned is excellent for traditional data processing and information systems. However, in real-time applications, timing becomes an additional, critical factor. Hatley, Mellor, and Ward have developed real-time extensions to the basic structured techniques.

For example, the design of an on-board flight computer may include specifications for programs that respond to wind-shear emergencies. These programs are triggered by analog sensors that detect such a condition.[1] EXCELERATOR/RTS allows you to document the timing and control aspects of such a system. Specific features that differ from base-EXCELERATOR include:

o   Data Flow Diagrams include dashed lines and circles to illustrate flow and processing of control information.

o   State Transition Diagrams, a new graph type, is added to depict states that a system can be in and the conditions and actions which cause a transition from one state to another.

---

[1]   This example was adapted from an example provided by Index Technology.

o   Matrix Graphs and Decision Tables, new graph types, are added to show processing control in a system.

o   Several dictionary entities contain new attributes that describe the physical implementation of a real-time system.

The remainder of EXCELERATOR/RTS mirrors the base product's capabilities. The only exception is the omission of data model diagrams as a graph type (although data modeling via entity-relationship diagrams, as taught in this tutorial, is retained).

Value-Added Products in EXCELERATOR's Product Line

Index Technology offers several add-on products and services. This section describes the products and services that are supported by InTech.

CUSTOMIZER: The "Have It Your Way" Product. *CUSTOMIZER* is an amazing product. It allows you to customize your EXCELERATOR environment to suit your own tastes and standards. You can literally use it, within specified restrictions, to create a custom product. Indeed, EXCELERATOR/RTS was created using *CUSTOMIZER*. *CUSTOMIZER* can be used to make the following type changes:

o   **Modify Product Structure and Integrate New Facilities.** You can change the menus and submenus, suppressing those facilities that you never use and integrating new facilities. The new facilities can link to non-InTech facilities (e.g. spreadsheets, code generators, fourth-generation languages, terminal emulators, etc.). For instance, you could customize the Main Menu, `SCREENS & REPORTS` to `PROTOTYPING`. Then you could add new facilities like thrid-party (non-InTech) fourth- and fifth-generation languages and applications generators to the four standard facilities provided by InTech.

o   **Create New Graph Types and Entity Types.** You can create new graph types, object shapes, and the like. For each graph type, you create an entity for the graph type and all of the object types on the graph. Consequently, as new methodologies and techniques become available in courses and books, you don't have to wait for InTech to add them to the base product! This commitment to making EXCELERATOR useful in any systems environment is highly unusual.

o   **Modify Existing Entity Types.** Within limitations, you can modify the attributes recorded for existing entity types. Consequently, you can customize EXCELERATOR's data dictionary description screens to require the attributes that your organization requires as standards or finds useful. This offers unparalleled flexibility for EXCELERATOR shops.

The price of CUSTOMIZER might unduly frighten off some potential users. However, because you need only one copy of CUSTOMIZER to legally customize all of your EXCELERATOR workstations, the true cost of CUSTOMIZER is the base cost divided by the number of workstations. Thus, the greater the number of workstations, the lower the true cost of CUSTOMIZER. Few data dictionaries can be customized to the degree of EXCELERATOR via CUSTOMIZER.

SPQR/20: A Professional Estimating Product. *SPQR/20* is a unique and highly valuable product. It can be thought of as an expert system for systems project estimation. The expertise is based on the authoritative work of Capers Jones, an internationally known consultant and author in the field of programmer productivity. *SPQR* stands for *Software Productivity, Quality, and Reliability.* The number *20* refers to 20 variables that effect the estimation models. The estimation models are based on a knowledge base of over 3000 projects from 200 organizations. After asking a series of questions about type of project, staff size, environmental factors, tools and techniques, etc., *SPQR/20* generates a forecast for your project. The outputs are summarized as follows:

o   Amount of source code that needs to be written (any of 30 languages or language classes)
o   Productivity gains that can be expected from the use of various tools and methodologies

o   Risks and the probability of said risks
o   Optimal schedules and overlapping options
o   Effort and dollar estimates for all activities
o   Staffing for activities
o   Maintenance costs for a five-year production lifetime
o   Quality and reliability estimates for the finished system

VDAM Mainframe Interface.   Available from InTech, VDAM allows you to store and share EXCELERATOR data (and other software packages' data) on IBM mainframes running the MVS operating system. VDAM acts transparently; therefore, there are no complex procedures to learn. Through VDAM, EXCELERATOR looks at mainframe- stored data as if it were just an alternative PC fixed disks.

Hotline (via Extended Maintenance Support Agreement). InTech maintains what is, in our opinion, one of the most responsive and helpful hotlines in the microcomputer industry. This support service is available to customers who are on the Extended Maintenance Support agreement. It should be pointed out that the Hotline provides technical assistance and product help, *not* consultation for methodologies and techniques.

XL GROUP: An Independent User Group for EXCELERATOR Customers. XL Group is an independent users' group for *EXCELERATOR* users. It pays to join such a group for several reasons:

o   You can more easily learn from others' EXCELERATOR experiences.
o   You can take advantage of a growing library of user-developed, value-added software. Although these tools are not supported by Intech, they can prove quite useful. For example, several EXCELERATOR users have developed mainframe data dictionary interfaces. Many organizations already have a considerable investment in mainframe data dictionaries. These are production dictionaries whereas EXCELERATOR is a development dictionary; therefore, the products do not really overlap functions (few organizations could afford to do EXCELERATOR-like work on their mainframes). Examples of interfaces include:

*   IMS DB/DC. This provides a two way link between EXCELERATOR and IBM's *Information Management System*, the workhorse DBMS in IBM's product line.
*   IDMS/IDD. This provides a two way link between EXCELERATOR and Cullinet's popular CODASYL standard database and dictionary.
*   Datamanager. This provides a two-way link between EXCELERATOR and MSP's popular general-purpose data dictionary.

Again, it is important to note that these tools are not supported by Index Technology.

o   The group sponsors an annual *EXCELERATOR User Conferences*. The 1986 conference drew over 300 attendees. Virtually everyone walked away having learned new ways to use CASE technology in their shops. Their is no substitute for this experience! XL Group membership is not mandatory for attendance.
o   You can funnel product and service improvement ideas through the User Product Committee.o You can keep up with the latest EXCELERATOR offerings.

Conclusion ... the Adventure Now Begins!

There you have it! You know enough about EXCELERATOR to really make it work in your career. We now suggest that you again read Lesson 1. In the context of your new skills, Lesson 1 should take on additional meaning. If your employer (or future employer) is not a CASE shop, you should be able to sell the concept. And if you are employed in an EXCELERATOR shop, you can now put it to use on a real project. Expect a learning curve. With experience, you will become a more efficient systems analyst - and your systems will be of a higher quality and better documented! Enjoy!

NOTES

INDEX